# Stopping the Next Pandemic

## HOW COVID-19 CAN HELP US SAVE HUMANITY

### DEBORA M

hachette
BOOKS

New York

*For James, Jessica, and Rebecca, who make everything possible.*
*And in grateful recognition of the scientists and journalists who do*
*their best to find out what's going on and try to save us from it.*

Hachette Books
Hachette Book Group
1290 Avenue of the Americas
New York, NY 10104
HachetteBooks.com
Twitter.com/HachetteBooks
Instagram.com/HachetteBooks

First trade paperback edition: September 2021

Previously published as *Covid-19: The Pandemic That Never Should Have Happened and How to Stop the Next One*

Published by Hachette Books, an imprint of Perseus Books, LLC, a subsidiary of Hachette Book Group, Inc. The Hachette Books name and logo is a trademark of the Hachette Book Group.

The Hachette Speakers Bureau provides a wide range of authors for speaking events.

To find out more, go to www.hachettespeakersbureau.com or call (866) 376-6591.

The publisher is not responsible for websites (or their content) that are not owned by the publisher.

Print book interior design by Six Red Marbles, Inc.

Library of Congress Control Number: 2020938926

ISBNs: 978-0-306-92424-8 (hardcover), 978-0-306-92423-1 (ebook),
         978-0-306-92422-4 (trade paperback)

Printed in the United States of America

LSC-C

Printing 1, 2021

# CONTENTS

# PREFACE TO THE REVISED AND EXPANDED EDITION

> We must never forget that a dangerous microbe anywhere in the world today could be everywhere in the world tomorrow.
>
> —Michael Osterholm and Mark Olshaker, *Deadliest Enemy: Our War Against Killer Germs*

In six weeks in March and April 2020, I wrote a book called *Covid-19: The Pandemic That Never Should Have Happened and How to Stop the Next One*. That's not a long time to write a book, especially during a fast-moving crisis.

But I could do it because a lot of my work as a journalist, over several decades, was about exactly what was, at that moment, finally happening: a pandemic of a new, "emerging" virus. And as I watched all those warnings come true, I realized that while people were glued to the day-to-day developments, most people were not reading about that larger story: how we were warned over and over again that this was likely, yet somehow most countries weren't ready—and what that means for the future.

That tale is still worth telling. To those who said, well, *this* book will be out of date pretty fast, let me say: this is the revised and updated version, and we have learned a lot, so I have added a lot, but hardly anything from before needed changing.

That's because what I wrote was mainly about things that would not change before you read this: what had happened already that led to Covid, and more important, that will lead to more outbreaks like it—the big picture of pandemic risk. Even in the early days, it was clear that Covid-19 both vindicated the warnings and showed what we must do to stop this—and worse—from happening again. That hasn't changed. If anything, it is even clearer now.

And people still need to hear it. Larry Brilliant, a leader of the campaign that eradicated smallpox in 1980, has a favorite saying. New viruses are increasingly jumping to people from animals, he says. "Outbreaks are inevitable. But pandemics are optional."[1]

That may seem an outrageous claim to make after all the tragic death and economic wreckage we've witnessed. But it is true. We can do our best to prevent outbreaks, catch those that happen anyway early, respond fast, and stop them from spreading out of control. We can stop the next pandemic.

Hence the title of this updated book. The whole point of understanding what happened to unleash Covid-19 is to stop something like this from happening again. And scientists agree, just as they did before Covid, that the question is not *if* that is going to happen, but *when*, and when could be any time. And as you will read in these pages, there are way worse viruses out there. Our civilization, like all the others before us, can collapse—and a bad enough pandemic could help do it.

So, please: learn about what happened with Covid. Use that information to stop the next one. Save humanity. That about sums it up.

There are some things I can say now that I could not say in 2020. One is about communication. People are not good at hearing warnings about the ill-defined future, but we are very good at listening when some threat is imminent or now underway. I decided to write this book mainly because I hoped that, now that a pandemic was actually happening, people would finally be open to all those pandemic warnings I had been hearing about from scientists for years. I still think that is true.

But fear and anxiety can make us a bit too open to some kinds of information. The World Health Organization (WHO) has talked from the beginning of this pandemic about an "infodemic" of misinformation that has accompanied it, and even a new science of "infodemiology" to track and predict misinformation the way epidemiologists track and predict viruses. Whether this will allow us to actually control epidemics of misinformation, conspiracy theories, and rampant denialism is an open question but an important one to answer if we are to stop the next pandemic.

Let's look at one example: masks. Surgical masks, as we all know now, were designed to stop germs getting out, so doctors wouldn't infect patients. Early in the pandemic, that seemed of little use: people with Covid were expected to be sick in bed, not walking around, exhaling virus. Meanwhile, years of observation had shown these masks don't totally protect you from taking in airborne germs. Plus masks were in short supply early on and needed for health care workers. So the advice was, don't wear them.

Then we discovered that around half the people with Covid had caught it from people who didn't have symptoms, didn't know they were infected, and in fact did walk around exhaling virus. This discovery in early 2020 was bad news, as it is so much harder to control a virus that spreads from people you don't know are sick, and recognition of the implications was slow as some medical authorities at first clung to denial.[2] But it was true, and it meant wearing masks made sense, even homemade cloth masks that were less effective than surgical masks.

So, unsurprisingly, our knowledge about a totally new virus changed. Hence, the advice changed: wear masks. Observation soon showed that masks stopped people from spreading virus and even, to some extent, from catching it.[3]

What is hard to understand about that? Yet many people brandished the fact that the advice about masks changed as proof doctors and scientists cannot be trusted, or claimed it was all a way to take away our freedom. Some still do. Perhaps it's just an eagerness to disbelieve bad news. But there are also plenty of people ready to foment and exploit mistrust, for whatever power—or just Facebook likes—it seems to offer. Infection rates soared in places where masks were rejected, and people died as a result.

Should experts have just explained it better? Or do we need to address the fear or discontent that drives people to believe, or become, conspiracy theorists? These are things we need to understand to ward off the next pandemic as surely as we need to watch the viruses you'll meet in Chapter 2 or develop the diagnostics I'll cover in Chapter 6. In Chapter 8, we'll look at a few things psychologists have discovered about people's reactions to disease, which helps explain the polarization that has afflicted some societies during the pandemic, in turn helping to drive the infodemic.

Another thing I can say now is how Covid fits into the even bigger picture of the risks we face as a civilization. Bear with me: this gets scary, but surprising reasons for hope and optimism are coming.

Every year, the Bulletin of the Atomic Scientists adjusts its iconic clock graphic, showing how close its experts think we are to midnight—doomsday—due to nuclear war, pandemics, climate change, cyber war, and other threats. In 2021, they left it where they put it the previous year, actually before the pandemic was official: 100 seconds from midnight, "as close to the end of humanity as the clock has ever been."[4]

The confused response to Covid, they wrote, made it clear how ill-prepared the world is to deal with global threats in general— yet humanity faces other risks "fundamentally greater," especially nuclear weapons and climate change. They called the infodemic a "threat multiplier" for all these risks. So is the growing economic inequality that also attended the virus: Covid-related unemployment is predicted to push 500 million into poverty, with its accompanying discontent and disease, by 2030.[5]

The nightmare combination, says the World Bank, is disease, war, and climate change, all feeding back on each other—the traditional horsemen of the apocalypse, as climate change means famine. Climate disruption leads to human migration and conflict, and the destruction of nature that promotes disease; conflict raises nuclear risk and displaces people, breeding disease and famine; disease causes poverty that slows progress on climate and breeds famine and conflict.[6]

We cannot talk about why this pandemic happened and how to

stop the next one without that bigger context. The good news is that addressing one of those threats helps with the others because they all come from the same place: the fact that we are now conducting our affairs, but not really governing ourselves, on a planetary scale.

There are lots of different, fine-grained ways to address pandemics, climate, and conflict. But the big story with all of them is that we all have to do it together, pooling resources, action, and information, on a global scale. As closed borders and less travel shrink our global interactions and pandemic hardships boost populist governments, that may not be getting easier.

So the reasons for hope and optimism I mentioned? Covid-19 has shown everyone how fragile our complex society is, as the virus disrupted support structures and caused cascading failures and economic damage everywhere. That may not sound like good news, but at least now no one can miss how profoundly united in vulnerability we are on this planet.

That is very important. The biggest thing stopping us from addressing many of these problems before was that many people couldn't imagine they could get as big or as bad as scientists said. Think of how nations, from China to Italy to the United States and everywhere in between, froze in denial or misplaced self-interest as Covid emerged, until their response was too late for many. Now think of how they are doing that with climate.

From now on—with luck—imagining the global upheavals that could befall us if we don't do something about the big risks should get easier. Imagining we can still live as separate tribes, each doing what we please regardless of its effect on the rest of this planet, should get harder.

Preface to the Revised and Expanded Edition

The operative word is *should*. In 2021, as genetic variants of the virus evolved and spread, it became clear that we won't be clear of Covid anywhere until people are immunized everywhere—yet we were failing to do that, as rich countries hogged the vaccine. Meanwhile, during the pandemic the wealth of the world's ten richest men increased by half a trillion dollars, more than enough to buy everyone a vaccine and ensure the pandemic pushed no one into poverty.[7]

More bad news—but, again, also a reason for hope. It means our society has within it the wealth and resources to do what we need to do. It's their distribution that needs adjusting, and our understanding that this is not a zero-sum game—that if one country tries to benefit at the expense of another, by keeping outbreaks secret or hoarding vaccine, we all hurt. For anyone prepared to look, the pandemic was a massive demonstration of that.

And yet too many people, even at the highest levels of leadership, still don't seem to understand what is happening. I still hear this called a "once in a century" pandemic, a temporary aberration after which we will return to "normalcy." It is not, whether you look back or forward.

Never mind that we had half a dozen actual pandemics in the hundred years before 2020, some pretty severe (HIV, anyone?). As we will see in these pages, the processes that lead to pandemics are continuing and accelerating, and we still don't have plans or tools to detect and respond to them before they spin out of control like Covid-19 did.

As I think we all know, the world was really, really not prepared for Covid. Some countries reacted well early on, but many others, as the Bulletin of the Atomic Scientists put it, "abdicated responsibility, ignored scientific advice, did not cooperate or communicate effectively." Soaring case numbers in countries that didn't contain the virus gave it chances to evolve, so it became even harder to stop for everyone. I really hope viral evolution has not sprung even nastier surprises by the time you read this.

Countries that failed to control the virus mostly seemed to fear the economic cost, even though economists quickly warned there was no real trade-off between lives and the economy.[8,9] You can't have a prosperous economy where unusually large numbers of people are sick and dying. A WHO panel reported in January that "governments willing to take the political and economic hit of harsh restrictions early in 2020" were then able to stop locking down sooner and were less economically damaged. "Trying to appease both public health demands and the libertarian views of the free market"—they singled out the United States, the UK, and Brazil—led to both huge death tolls and flailing economies.[10]

The real problem, they stressed, was that these countries were not just making their own mistakes—poor pandemic management in one place spawned problems, economic or viral, that then hurt people everywhere. The *British Medical Journal* and the US Council on Foreign Relations both proposed dealing with this by making public health malpractice a crime against humanity so negligent governments could be held accountable—and might try harder next time.[11,12]

Because the malpractice was enormous. Country after country cut case numbers with lockdown, then failed to do the testing, tracing, and isolation that would have kept them down, or relaxed restrictions too early, so the numbers soared again, and repeated lockdowns deepened the pain. In 2020, a senior UN official said to me that he didn't know what preparing for something like this would even look like. Easy: you prepare the plans and resources you need to stop it ever getting like this. That this is possible was abundantly demonstrated by the countries that did a good job of managing their pandemic. We need to learn why we didn't all do that this time and make those preparations before there is a next time. That's what a lot of this book is about.

It might help to realize that a swarm of well-established psychological biases keep people, especially in groups, from acting on warnings of disaster. We all believe what confirms our own biases (*we* can't die of a pandemic) and ignore inconvenient truths (actually, we can). We all have a bias toward optimism, except maybe scientists and journalists like me who have spent a lot of time talking to scientists. Research shows that virtually no one who hasn't been schooled in it understands what "exponential" means, and therefore few understand why you need to respond to an outbreak fast.

That's when normality bias kicks in: we assume nothing bad is happening if we are surrounded by people who don't look worried. To paraphrase Rudyard Kipling, if you can lose your head when no one around you is losing theirs, you should maybe consider a career in public health.

Then there's "othering," the tendency not to worry about things that seem to affect only people seen as foreign, or poor, or racially different. In the early days of Covid, Western governments seemed convinced the epidemic that started in China could not possibly come "here." Meanwhile politicians have few incentives to spend on risks that may or may not strike during their term of office. And if disease does strike, governments everywhere seem compelled to hide the truth from the public "to avoid panic," when what we need most is to be trusted with the facts. Both facts and trust matter.

But again, the good news: Covid has given us a chance to recognize and maybe fix all that, while we are still living with facts no one can argue with. Psychologists say that if we recognize these biases, we can communicate plans to people in ways they understand and accept. And Covid revealed one enormous cause for hope: individually and in groups, countless people responded selflessly and tirelessly to help others and fight the virus despite their own pandemic hardships. We need to learn to tap those deep wells of human resilience to prevent the next pandemic. After all, adequate preparedness for this pandemic would, it has been estimated, have been five-hundred-fold less than the cost of a pandemic in dollars, inestimably less if accounted in lives.

Especially if it's a worse one than Covid-19. As Mike Ryan, head of emergency response at the WHO, said in late 2020, "This pandemic is very severe, but it's not necessarily The Big One."

We need dedicated institutions whose job it is to watch the world's viruses, decide what needs preparation or response, and see that it gets done. We must stop dangerous farming and industrial practices that breed new human infections. People who do not see our natural planet as worth conserving in its own right may now see another point: destroying nature—and that includes climate change—leads to dangerous encounters between animals and humans that can cause deadly, destabilizing, very expensive pandemics.

The effort will have to be truly multinational. Pandemics hurt everyone: we cannot continue to risk everything on individual countries' supposedly inviolable rights to act however they want. We all share the risk; we must all share responsibility for managing that risk. As complexity scientists have found, it takes a network of players to manage a complex system. We need to start networking better. And more than anything, we need to build trust among countries and people so that we all tell everyone when a disease is emerging—fast.

Our job now is to take the political momentum created by Covid-19 and put global institutions in place to achieve that, and whatever else we need to save ourselves from the threat of pandemic disease. The WHO plans an international meeting in November 2021 to discuss a new pandemic treaty that could help a lot, but we will need more than that, scientifically, socially, politically. We will consider some of the options in these pages. The effort might even be a good model for the other existential problems we face.

Meanwhile, the Covid-19 virus is likely to remain with humanity, but experts believe the crisis will end. Vaccination and rapid tests will tame its biological and economic impact, even as the virus itself remains in circulation. Even that may be no bad thing, as long as the virus does not evolve to become significantly more deadly, especially in the very young. Ultimately, if we all start encountering and surviving it early in life, Covid-19 could become a mild disease of childhood, so we all grow up with at least partial immunity—and it will join its coronavirus relatives as a common cold.

But the complacency that led to this pandemic must not return, or we will certainly have more and worse ones. We have had our collective faces rubbed in our collective fragility by a humble bat virus. Let us, please, learn our lesson. I hope what I have written here helps lead you to further understanding—and action.

It is not too late to save ourselves and our children. In fact, the upheaval of the pandemic might be just the creative destruction we needed to finally get serious about it. The fight will need all of us, and it will start with understanding what just happened. Let's get started.

# PREFACE

In November 2019, a coronavirus from a common little bat jumped, somehow, to a human, or maybe a few of them. It just happened that the virus could spread easily among people already, or it evolved fast, as these viruses can. By December, a cluster of people were hospitalized with severe pneumonia in Wuhan, China, and it wasn't the flu.

Not enough was done to contain this new virus until weeks later, on January 20, when China told the world it was contagious. By then there were already so many cases in Wuhan, the city had to be locked down three days later to control the epidemic—but it had long since spread all over China and to other countries. The virus was named SARS-CoV-2 because it was so similar to another one we had barely managed to beat back in 2003. As you know, the disease it causes was named Covid-19: "co" for corona, "vi" for virus, "d" for disease, and 19 for the year it appeared. A lot of people just call it the coronavirus.

Three months after Wuhan was locked down, some two billion people worldwide were also in some form of lockdown, and everyone, everywhere faced infection with the virus, with few effective treatments and no prospect of a vaccine anytime soon.

I wrote the preceding words in March 2020, when the reality of the Covid-19 pandemic was just beginning to sink in. Now, in early 2021, none of those basic facts have changed. We know a vast amount more about how to respond to a viral pandemic, and even more, how not to. We know how to make a few vaccines now.

But now, as then, the bottom line is that Covid-19 has infected the entire human world. This pandemic has been like a big dog picking up our fragile, complex society in its teeth and shaking it. Lots of us have died, officially more than three and a half million, but probably more than eight million, by mid-2021. But scientists estimate that, based on the ages of the people who died, the world has lost more than twenty million years of human life—an average of sixteen years denied each person who died. Younger people were less likely than older ones to die of Covid but had more life expectancy to lose—so 45 percent of those lost years were from people between fifty-five and seventy-five, whereas 30 percent were from those younger than fifty-five. Depending on the country, that was up to nine times more than the years lost to the yearly flu epidemic, even though flu takes more years per death by killing young children as well as the old.[1]

And those numbers will grow, as lots of us continue to die, either from the virus itself or from the long-term poverty, political and economic dislocation, and overloaded medical systems that will be the pandemic's legacy. Some aspects of our society will change for the worse, some maybe for the better—but, either way, forever.

And through it all, we have been deluged with reams of news

reports and instant analyses, heartbreaking frontline accounts, revised government instructions, and new medical advice, plus probably the most staggering global outpouring of instant scientific research in history, all trying to predict what's coming next and figure out how to mitigate this disease disaster.

But you know all that.

And still there is the question: How could this happen? This is the twenty-first century. In much of the world, we have wonder drugs and flush toilets and computers and international cooperation. We don't die of pestilence anymore.

Sadly, as we all know now, yes, we do. But what remains especially sad for a science journalist like me who writes about disease for a living is that this pandemic was not exactly a surprise. Scientists have been warning for decades, with mounting urgency, that this was going to happen. And journalists like me have been relaying their warnings that a pandemic is coming and that we aren't prepared.

How did we find ourselves in this situation? In short, there are more and more people, and too many of them have had to put ever-increasing pressure on natural systems to get the food and jobs and living space they need. That means pushing into wilderness that harbors new infections and intensifying food production in ways that can breed disease. Covid-19, Ebola, and worse come from destroying forests. Worrying flu strains and antibiotic-resistant bacteria come from livestock. Yet we have neglected to invest in the things that discourage infectious disease: public health, decent jobs and housing, education, sanitation.

Then the impact of the new pathogens we unearth is magnified by our ever-increasing global connectedness, as we crowd into cities and trade and travel in an ever-denser global network of contact. So once public health fails and contagion appears anywhere, it goes everywhere. We know so much about beating disease, yet fragmented governing structures, lack of global accountability, and persistent poverty in so many places ensure that those failures happen and disease propagates.

Despite all that, we know what we need: much better understanding of potentially pandemic infections, fast detection of new outbreaks, and ways to respond to them quickly. I'll be looking at that in this book. So far, we haven't been able to do that effectively, where and when it is most needed, before the pandemic or even when the first small outbreak strikes.

Starting in 2015, two labs—one Chinese, one American—published their investigations into the tribe of bat viruses that is the source of Covid-19. They immediately recognized the threat. One lab called them "pre-pandemic" in a prominent biomedical research journal and highlighted their "potential for human emergence" in the title of the article so no one could miss it.[2] The other group wrote in their title that the viruses were "poised for human emergence" and called them "a substantial global threat to public health."

The labs themselves kept studying the viruses, but other than that, nothing was done by any public health authorities to prepare for an outbreak. We could have learned far more about them, designed some vaccines, looked into tests and treatment, studied ways these viruses (or the bats that carried them) might infect human populations—and shut those down. None of that happened. It was no one's job to take on those tasks with this kind of threat, even when it materialized.

Yet we needed so much to be in place if one of these viruses went global—which one did. You don't need to be told. Testing. Ventilators. Drugs. Vaccines. PPE (personal protective equipment) for doctors and nurses. A plan for using old-fashioned quarantine and isolation to stop this kind of virus from spreading. A plan for dealing with the economic impact. Measures to contain the virus as soon as it emerged in a population anywhere, so we might not even need those things. Experts and governments have been talking intensively about pandemic preparation for nearly two decades, and still we weren't prepared.

And as we noted, this kind of virus wasn't—and isn't—even the only viral threat out there, yet we're just as unprepared for the others. I wrote the following for *New Scientist* magazine in 2013, the year the Covid-like viruses were discovered, about a visit to the World Health Organization's then-shiny-new situation room, and what might happen if H7N9 bird flu, the virus causing concern at the time, went pandemic:

As it stands, the World Health Organization's top brass will watch any H7N9 pandemic unfold from their strategic operations centre. Information will flood in; body counts will mount. Governments will be told that their demands for vaccines and drugs cannot be met. They will issue declarations, hold briefings, organise research, tell people to wash their hands and stay home. Mostly, though, they will just watch helplessly.[3]

Sound familiar? Especially the part about washing your hands and staying home?

## Preface

I don't claim to be prophetic: I'm not. Other journalists and scientists have said as much and more. As far back as 1992, the top infectious disease scientists in the United States warned about "emerging infections," declaring that the threat from "disease-causing microbes . . . will continue, and may even intensify in coming years."[4] If that sounds like unusually cautious language, even from scientists, it's because they were afraid any stronger language would trigger disbelief. That's almost all that has changed.

It's not that they weren't heard. In the years since then, we all started half-expecting a pandemic. Pandemics became part of the cultural background noise, reflected, with varying balances of science and entertainment (and zombies), in films like *Outbreak*, *Contagion*, and *I Am Legend*. There was some disease surveillance set up, new international rules written, a lot of virus research. A few countries had pandemic plans, on paper. Yet when the lockdowns began, in many places toilet paper was in more demand.

The only real surprise when Covid-19 finally hit was the sheer extent to which most governments simply had not listened to the warnings. We were unable as a planet to muster our considerable scientific understanding of disease in time to soften the blow, never mind preventing it in the first place. And, as I will explain in the coming pages, we could have—at least a lot more than we did. Science didn't actually fail us. The ability of governments to act on it, together, did.

Experts had warned about the lack of preparation in addition to the risk of a pandemic itself. The few countries with pandemic plans built them around a very different virus, flu, and regardless, many failed to stockpile or acquire the most basic essentials for making the plans work. I'm not sure their response would have been much more effective if this had been a flu pandemic. Which we will also have at some point.

The WHO made it very clear how to contain Covid-19, but few countries followed their advice entirely. A few did, and they showed what should have been possible for all countries. The rest did pick-and-choose variations on the WHO's advice and/or that of their scientific or political advisers. Nearly all countries were more or less too late to limit the damage as much as they might have, and the pain of lockdowns and economic dislocation in some places seemed to rival that of the disease.

But you know that.

So, besides the question of how this could happen, the other big questions are, Can this happen again? And can we do better next time? The answer to both is yes. Some real pandemic planning is now in order, because the Covid-19 pandemic may not even be the worst we could see. And even Covid-19 could still have some tricks up its tiny sleeves.

First, let's look at the immediate future from the virus's perspective. Eventually, after considerable death and disruption, most people in the world will have been exposed to or vaccinated against Covid-19 and will be, we hope, immune to further infection with the same virus as a result, at least temporarily.

So, with fewer people around who it can still infect, new cases should slow to a trickle. The virus might even quietly die out, as its sister virus SARS did in 2003 when we blocked enough chances for it to spread. It must be said, though, that few epidemiologists or virologists really expect that, especially now that we have seen how the SARS-2 virus that causes Covid can mutate to escape our immunity and keep going.

More likely is that it will adapt to its new situation. RNA viruses like this one can evolve quickly, and although coronaviruses aren't quite as volatile as some, several widespread genetic variants that help the virus spread more easily—which you would expect, as it is what benefits the virus most, so that is what evolves—arose within the pandemic's first year.

If most of us have long-lasting immunity to the virus, it might start pouncing on new, susceptible humans, becoming another disease of childhood. Or, like flu, it might continually evolve to evade the immune defenses our bodies mount in response to exposure or vaccines and continue its global rampage, perhaps a bit less deadly this time—or perhaps a bit more. The comforting story that viruses become more benign as they adapt to us is simply not true and was blown out of the water by SARS-2 in its first year. It all depends on what works for the virus, and it can go either way. We will look at that later in the book.

This pandemic has moved fast since it started. You may already know more than I do now about which of those scenarios is playing out. There aren't, broadly speaking, a lot of different things a disease can do, bound by the implacably quantitative laws of epidemiology, the science of epidemics we all learned so much about during 2020.

Until then, horrific as it has sometimes been, we can be grateful it hasn't been worse. At least up until the time I am writing this, Covid-19 does not have a massive death rate—best guesses are around 1 percent of cases, some ten times more deadly than ordinary winter flu. SARS was ten times deadlier than that. Fortunately,

## Preface

SARS never learned to spread like Covid-19—and, with luck, Covid-19 will never learn to kill like SARS. Think about what this pandemic would be like with ten times the death rate.

And as many of us have painfully learned, Covid-19 mostly kills older people. Speaking as one myself, I don't wish to be cavalier about this, but the brutal fact is that losing people in old age does not cause as much economic or social disruption as losing children or people of working and childbearing age. And even that will pass: eventually we should have enough vaccines, and perhaps antiviral drugs, to protect everyone, including the elderly.

So why write a book about this when there's still a lot we don't know? People asked me that when I wrote the first edition of this back in spring 2020; one can still ask that now as I update it for 2021. The answer is the same: because we already know enough to say some important things, and we need to do that while memories of these hard times are raw enough for people to hear them.

The first thing to say is that this was predicted and could have been, to a large extent, prevented.

As for prediction, I am just one of many journalists who have been warning about the threat of a pandemic since the 1990s—and some were at it earlier. Since at least 2008, the US Director of National Intelligence has warned the president that a pandemic of a virulent, novel respiratory virus was the most serious threat the country faced. In 2014, the World Bank and the Organization for Economic Co-operation and Development (OECD), the club of rich nations, called a pandemic the top catastrophic risk, outweighing terrorism. Bill Gates has been warning for years that we aren't ready for a pandemic.

Second, this pandemic won't be the last one. There are simply too many potentially pandemic germs out there to predict which will emerge next. But before Covid-19 happened, we knew coronaviruses were among the leading possibilities: they were on a WHO watch list. Even with such warnings, we didn't do enough preparatory work on drugs and vaccines for coronaviruses like Covid-19 to allow us to easily adapt and produce them now—and we still haven't for many other viruses that pose a threat, including some real nightmares we will meet in Chapter 2. We need to do that now.

We also need to do some serious pandemic planning for when the next one happens. The Center for Health Security at the Johns Hopkins Bloomberg School of Public Health was among the institutions already trying to do that before Covid. Among other efforts, they were running computer simulations of hypothetical pandemics as a training exercise for public officials. A month before the first clusters of unusual pneumonia were recognized in Wuhan, they ran one called Event 201, starring a fictitious virus that was nearly a dead ringer for Covid-19. I can think of few better illustrations of how we knew this was coming.

Let me emphasize that this was a total coincidence: this was a what-if scenario playing out in a computer model of US society, featuring a made-up virus. They chose a coronavirus for the simulation because we knew these viruses were a threat, and also to show how disruptive even a relatively mild virus can be.

They succeeded. The result of the simulation was what we started living out just a few months later: overwhelmed health care, disrupted global supply chains, needless death, economic dislocation. And a table full of officials from government and industry sitting there, saying, *If this were to happen, there's not much my sector/department/office could do.* Turned out many of them were right.

And the people who wrote that simulation were going easy on the officials—maybe so they'd sit through the entire afternoon and not be so horrified that they'd quietly slip out at the coffee break, trying to forget what they'd seen. There are much worse viruses out there that could trigger a pandemic that would kill more people at younger ages.

It will not be much comfort to those who have lost or will lose loved ones to Covid-19, but so far, believe it or not, we've been lucky. In addition, what almost no one realized before Covid-19 happened—I don't know how many realize it now—was what a pandemic could do to our complex, just-in-time society, and that economic domino effects would cascade through our tightly coupled global support networks.

What we need to remember, though, is that we *will* have another pandemic. And it could be worse.

So we have to do better—and we can. The hard-earned good news is that Covid-19 has shown us what we need to do. We cannot let a virus catch our interconnected global community this stupidly flat-footed again. We cannot let it break those interconnections either, at least not all of them. If this pandemic teaches us anything, it is that up against a contagious disease, we are all in this together. One big early lesson was that no country can really seal off their borders anymore or go it alone. Our society is global; our risk is global; our response and our cooperation must be global.

I can't think of a time when this pandemic will be "over" enough to provide a better vantage point for looking at these things. When the virus does grind to a halt, or we tame it with vaccines, it seems all too likely that we will drift back into a status quo of spending on wars and weapons—and on recovering from the economic damage Covid-19 did—not on preparing for the next virus. We will need to forget this nightmare, and to judge from past pandemics, we will.

Yet at this moment, despite our pandemic fatigue, the subject has our sadder but wiser attention. We can already say a bit about how this happened, and why, and what our options are to start doing better. Many scientists know this, and governments, we hope, will learn. But a lot of other people need to think about this, too, whatever you do in life, in the kind of detail that will allow you to help make the changes we need.

In any disease emergency, certainly in a pandemic, it is vitally important to tell everyone the whole truth—what we know and what we cannot know—and not hold back for fear of scaring people. That is a mistake that governments and other authorities made repeatedly with Covid, and make all too often with bad news stories like diseases.

What is happening might be scary, but saying so might galvanize people to take more effective action. Sometimes fear is necessary. That's why we have it.

But it shouldn't have to come to that. This is where you come in. Learning from this pandemic and preventing the next one will take political action of all kinds, from everyone.

The more people understand what we need to do, the more likely it is to be done. People vote. People march. People pressure. People decide to study virology or public health or nursing or supply chain management or communications. Public activism drove the development of HIV drugs and made them affordable. It drove the introduction of sanitation, the massive success of vaccination, the beginning of the end of smoking.

We can do it again. We have to.

To find out what is happening with Covid-19 right now, read the news. For exposés and analyses of what this and that government or politician did wrong in dealing with it, also read the news and the stories that will pour out over the coming years.

In this book, I'm going to give you the big picture. We'll take an in-depth look at what happened and whether we could have stopped it, before looking at the recent past to learn the natural history of some of the more amazing natural phenomena that make us deathly ill. We'll see how previous pandemics and threats of pandemics should have prepared us and learn the lessons we failed to apply before and after Covid-19 emerged. Then we can talk about what we need to do better before the next one hits.

I hope that we will eventually do more than talk.

CHAPTER 1

# Could We Have Stopped This Whole Thing at the Start?

> Every disaster movie starts with someone
> ignoring a scientist.
>
> —popular poster at the April 2017
> March for Science

So how did we end up with the Covid-19 pandemic? Could we have stopped it once it started? Could we have stopped it from starting at all?

If your house burns down, you ask two things. First, how did a fire get started in the house to begin with? Second, and most urgently, given that it did—and we saw it happen—why didn't we put it out before it spread? We'll look at the first question later in the book. Let's look at the second now. What happened to unleash a Covid-19 pandemic on the world?

The first inkling I, like many others, had of the gathering storm that became Covid-19 was a post on the online forum ProMED.

The machine-translated report from Finance Sina, a Chinese online news site, read,

> On the evening of [30 Dec 2019], an "urgent notice on the treatment of pneumonia of unknown cause" was issued, which was widely distributed on the Internet by the red-headed document of the Medical Administration and Medical Administration of Wuhan Municipal Health Committee.[1]

It was December 31, and in our suburban French village, just over the Swiss border from Geneva, the sun was coming up. I had family in for the holidays and had solemnly promised to stop working.

But, I told myself, that didn't mean I couldn't take a peek at ProMED, just to make sure I didn't miss anything important.

ProMED—the Program for Monitoring Emerging Diseases of the International Society for Infectious Diseases, a scientists' organization, formally called ProMED-mail—is the world's leading online reporting system for new or emerging infectious disease. Despite its importance, it's a nonprofit run mostly by volunteers, on a shoestring, with grants and donations. It was set up in 1994 as infectious disease specialists shaken by the emergence of AIDS in the 1980s uneasily realized that other new diseases might be out there, and that we needed an early warning system.

It consists of moderated daily reports of worrying medical events from contributors everywhere: doctors, vets, farmers, researchers, ordinary citizens, even agriculture labs (crops get diseases, too). It's all in understated sans serif plain text—old-fashioned Helvetica, direct and to the point like the scientists who mostly read and contribute to it. Everything is classified by disease and place and date. The moderators, most of them veterans in their areas, tell you what they make of the reports, and I often cut straight to their comments. ProMED is one of the things humanity did right to prepare for disease emergencies like Covid-19.

For disease researchers, public health people, and science reporters like me—as well as anyone fascinated by the daily reality show—ProMED is required reading. When I ducked into my office that day, hoping it was early enough that my family wouldn't notice, the giant Sina Corp's financial bulletin was reporting people with severe, undiagnosed pneumonia in the central Chinese city of Wuhan, in Hubei province.

Many had connections to a seafood market. There were already twenty-seven cases.

A red-topped bulletin—rendered "red-headed" by the machine translation—must be an emergency alert, I guessed. The reporter from Finance Sina had verified it by calling the official hotline of Wuhan's Municipal Health Committee the next morning. It was true. The story went out.

And it was worrying enough to make someone send it to ProMED. It wasn't hard to see why.

Pneumonia is not a disease caused by a specific germ, like measles or flu. It just means any infection that inflames your deep lungs, the part with the air sacs called alveoli. Those sacs are what your lungs are all about: you suck air into them, and oxygen pours across the alveoli membranes into the oxygen-starved blood on the other side. The carbon dioxide waste in that blood meanwhile pours into the alveoli, and you exhale it.

3

If those delicate membranes are damaged by an infection, they can start leaking fluid, and the sacs can fill up. That stops oxygen from getting to the membranes and entering your blood. If this gets bad enough, you effectively drown in your own fluids.

A respiratory infection—be it virus, bacteria, or fungi—may invade your nose, throat, or the deeper, bronchial air passages and give you a cold or a bad cough. But if it gets into the alveoli, that's pneumonia, and it can kill you.

The fact that this pneumonia was undiagnosed was the red flag that got ProMED's attention. Normally, white blood cells defend your alveoli from the bacteria that are always there, pulled in by the billions on every breath. Winter flu viruses knock out this key part of our immune system, and then the bacteria can grow, causing pneumonia. Therefore, most winter pneumonia is first treated with antibiotics, which kill bacteria. In Wuhan, this apparently wasn't helping. Nor, presumably, were diagnostic tests for flu or the other usual suspects.

The Municipal Health Commission was holding a special meeting, said the report. But they made a point of saying they thought it wasn't the SARS virus. SARS emerged in China in 2002 and rampaged through twenty-seven countries in 2003, causing severe pneumonia and killing 774 people.

Good, I remember thinking. SARS may not get talked about much anymore outside the countries that were affected, except by us disease buffs, but it was vicious, packing a 10 percent death rate. It was stamped out with an enormous international effort—and luck—with only the classic techniques of isolation and quarantine, mainly because it was clumsy at spreading among people. But if this new illness wasn't SARS, what was it?

4

The market connection was worrying. A seafood market in China is also a "wet" market that sells live animals, and many vendors sell exotic, wild creatures. The SARS virus came from bats and is thought to have jumped to people in a wet market.

Granted, there have been other reports like this on ProMED. In 2013, there was undiagnosed viral pneumonia in health care workers in Anhui province in China. In 2006, people in Hong Kong had undiagnosed pneumonia after visiting several parts of mainland China.[2] The ProMED moderator asked for more information in both cases, but further posts never appeared, so presumably no notable disease resulted.

This time, though, there was a worrying comment by Marjorie Pollack at the bottom of the post. Pollack is a doctor and epidemiologist, a thirty-year veteran of the US Centers for Disease Control and Prevention (CDC), and the doyenne of ProMED's international team of moderators. She was involved in one of its proudest moments: alerting the world, on February 10, 2003, to the mystery pneumonia in Guangdong, later named SARS, nearly two months before China opened up about it.

What she wrote that holiday morning gave me that queasiness you get when you're trying very hard to dismiss a feeling of foreboding. Besides the news report, she observed, there was a lot of online comment about this.

Twitter and its Chinese counterpart, Weibo, weren't around when SARS broke out, but online chatrooms were. "The type of social media activity that is now surrounding this event is very reminiscent of the original 'rumors' that accompanied the SARS-CoV outbreak," wrote Pollack. "More information on this outbreak . . . would be greatly appreciated. And," she added hopefully, "if results of testing are released."

5

What was different from SARS, she noted, was the transparency of the Chinese authorities. In February 2003, Chinese officials discouraged press reports on the undiagnosed pneumonia and did not immediately report it to the World Health Organization (WHO).[3] It didn't start fully reporting cases until that April, by which time SARS was spreading across China and East Asia and in Canada.

In the ensuing eighteen years, there has been an astonishing revolution in China's politics and prosperity, so this new outbreak was occurring under very different circumstances. Nonetheless, after picking up the same bulletin from Wuhan as ProMED on the 31st, the WHO had to ask about it twice—on January 1 and 2— before Chinese authorities replied on January 3.

China said there were forty-four cases of pneumonia of unknown cause in Wuhan, some severe, and some linked to a fish market, which had been closed on January 1.[4] It later emerged that the first cases probably occurred in November or even earlier, but a few respiratory infections during flu season had not attracted attention until hospitals started getting unusual clusters of severe cases in December.[5] Meanwhile, the new information from China just made my queasiness worse. I emailed the ProMED report about it to my news editor at *New Scientist* on Saturday, January 4, so I'd catch her when business resumed that Monday, with the comment: "Happy New Year. Hope the hols were restful. This is looking very worrying, IMHO (and I remember SARS . . . )."

There were worrying reports that people had been arrested for discussing online whether the mystery pneumonia might be a resurgence of SARS. Hubei authorities were cited as saying this was not true, as "no person-to-person transmission has been found so far."[6]

That became a recurring theme. On January 3, China assured the WHO that there was "no evidence of significant human-to-human transmission."[7] On the 8th, ProMED reported that China's Center for Disease Control and Prevention (China CDC) had identified the infection as a coronavirus, the same family of viruses as SARS, but repeated that there was no person-to-person spread.[8]

I hadn't planned to go back to work yet, but I wondered if I should start looking into this story. It seemed unlikely to be significant without person-to-person spread. Animal viruses sometimes manage to jump to people, and even kill them, but fail to transmit between humans, like the notorious H5N1 bird flu. Without that, this outbreak might just fade, I thought hopefully.

But on ProMED, Pollack was sounding increasingly suspicious. So was Jeremy Farrar, the head of the medical research foundation the Wellcome Trust, and before that, head of Oxford University's medical research lab in Vietnam, where he dealt with SARS and H5N1 imported from China. On January 10, he tweeted that if "critical public health information is not being shared immediately with @WHO—something is very wrong."[9]

Something was. What follows is what I have pieced together from apparently reliable press accounts that had emerged up to early 2021.

Central to the story is the virus's gene sequence, the precise order of the four smaller chemicals that are strung like beads along a strand of DNA or RNA, making up the genes that code for different proteins. Human genes, the recipes for our proteins, are made from DNA; the virus that causes Covid-19 keeps its genes as the slightly different, less stable RNA.

Determining the sequence of those genes allows us to identify the virus causing a disease and design precise tests and vaccines for it. China is a world leader in gene sequencing, and it has numerous private sequencing firms that work with hospitals to diagnose infections.

On February 26, the investigative Chinese business news outlet Caixin reported that at least nine samples of the coronavirus in Wuhan had been fully sequenced by the end of December 2019— and the sequences showed it was worryingly similar to SARS, a virus that did spread person to person. At that point researchers could readily have used those sequences to make specific tests for the virus, which could have revealed the scale of the problem— that mild and asymptomatic cases had already spread farther than the hospitalized cases in Wuhan suggested. Other countries— investigations have shown the virus was in Italy, France, the United States, and elsewhere in December—could have tested and started to contain it.

But no sequences were made publicly available until January 11, and China insisted until January 20 that, unlike SARS, the virus did not spread person to person. The delay prevented earlier action to stop the virus going pandemic.

On December 24, Caixin reports, Wuhan Central Hospital sent a lung sample from a sixty-five-year-old employee of the Huanan Seafood Market, who was severely ill with pneumonia, to a sequencing firm in Guangzhou. Three days later the results came in, and the firm urgently phoned the hospital to tell them it was a previously unknown coronavirus.

It also told the Chinese Academy of Medical Sciences and sent its top scientists to Wuhan to discuss the results. Everything was kept confidential.[10]

In their later publication of the results, the firm reported that the virus was 79 percent identical to SARS and 87.6 percent identical to a previously published sequence of another bat coronavirus. But the bit of the virus that binds to cells to infect them was much more similar to SARS. That suggested this virus, like SARS, could use that binding site to spread between people: accordingly, the sequencing firm disinfected its equipment, and the patient was isolated. He later died.[11]

Then Wuhan Central sent another sample, from a forty-one-year-old pneumonia patient with no link to the market, to another sequencing company, this time in Beijing. And that company, Caixin reported, made "a small mistake."

On March 11, Dr. Ai Fen, head of the emergency department at Wuhan Central, told the Chinese magazine *Renwu* (*People*) that on December 30, she saw the sequencing result from the Beijing company. It read "SARS coronavirus."[12]

It's very possible a partial sequence would have identified the new, then-unknown coronavirus as SARS, since many of their genes are similar. In fact, the official committee of virologists tasked with naming the new virus decreed on March 2 that the two are the same species.

They renamed the SARS virus SARS-CoV-1, "CoV" for coronavirus. The virus that causes Covid-19 officially became SARS-CoV-2, like the second installment in a movie franchise: *SARS 2: This Time It's Everywhere.*

But that December Dr. Ai didn't know any of that. She told *Renwu* that the diagnostic report made her break out in a cold sweat. SARS was a nightmare for China, officially infecting 5,327 people and killing 349, many of them doctors and nurses infected while caring for patients.

Ai took a picture of the report on her phone, with "SARS coronavirus" circled, and sent it to other doctors in Wuhan, including an ophthalmologist, Li Wenliang. He passed a warning on to medical students about wearing protective gear around patients with pneumonia quarantined in the emergency department. The news spread fast: the hashtag #WuhanSARS started trending on Weibo, China's substitute for the banned Twitter.[13]

The hospital told Ai that night not to spread information about the pneumonia cases, so as not to cause panic and "damage stability." The hospital's disciplinary committee reprimanded her.

Staff were told not to exchange messages about the disease, Ai told *Renwu*, and, astonishingly, not even to wear protective masks and gowns, for fear of causing alarm. After all, there is no need for such protection with a virus that is not supposed to spread between people.

The Japanese newspaper *Mainichi* echoes Ai's story. In late January, it reported that at 1:30 a.m. on December 31—the night of Ai's test result—eight doctors in a group chat discussing the worrying news were summoned by authorities and told to write self-critical essays about spreading rumors.[14]

They did. The crackdown silenced doctors. And that day, researchers at the University of Toronto discovered, content relating to Wuhan and pneumonia started being censored on the popular messaging and livestreaming platforms WeChat and YY.[15] "If I had known what was to happen, I would not have cared about the reprimand. I would have fucking talked about it to whoever, wherever I could," Ai told *Renwu* in a translation carried by *The Guardian*.[16]

As the epidemic worsened, Li Wenliang was hailed by the government in Beijing as a whistleblower. On February 7, he died of Covid-19. "I am not a whistleblower," Ai modestly told *Renwu*. "I am the one who provided the whistle."

The speed with which Wuhan authorities cracked down on any discussion of the outbreak suggests that a decision had already been made, either at provincial or national level, to keep the outbreak quiet. That means authorities were worried—which seems unlikely if they really believed the new virus was not contagious.

Caixin reports that the Guangzhou sequencing firm later double-checked the sample analyzed in Beijing and found it indeed contained the same virus they had discovered, not SARS. Yet they say the mistaken diagnosis was a lucky accident. It "directly caught the attention of doctors in Wuhan and sounded a warning to the public through social media, and to some extent saved many lives," they wrote—because now the secret was at least partly out.

Meanwhile, doctors were realizing that the new virus was in fact contagious. Zhang Jixian, head of respiratory and critical care at Hubei Provincial Hospital, told reporters in February that she knew on December 26 when an elderly couple came in with pneumonia. On a hunch, she called in their adult son, even though he had no symptoms.

In all three of them, CT scans revealed the "ground glass" opaque areas in the lungs typical of the new disease. Three cases in one family suggested the virus did, in fact, spread between people. The son later developed Covid and recovered.

Zhang bought her staff protective clothing to wear under their normal white uniforms and made them wear protective N95 masks. She also, says Caixin, insisted that the hospital report the cases to the local branch of the China CDC. The result was the Wuhan health department's emergency notice on December 30 that was sent to ProMED.

And there it got Marjorie Pollack's attention, partly because of the social media buzz that accompanied it, which was so like what happened with SARS seventeen years before. It is ironic that the buzz was itself the result of the misdiagnosis of Covid as SARS.

But now the world was watching. It was about to get something important to look at.

Also on December 30, back at Wuhan Central, doctors took a fluid sample from the lungs of another pneumonia-stricken employee of the seafood market. On January 3, they packed some of it in dry ice to be hand-carried via high-speed train to Zhang Yongzhen of the Public Health Clinical Center at Fudan University in the coastal metropolis of Shanghai.

What happened next got Zhang listed a year later as one of 2020's top ten scientists in the prestigious journal *Nature*. Frankly, I'd have used a smaller number.

Wuhan Central had been collaborating for five years with the Shanghai lab in a project to track what viruses were circulating in China. This is exactly the kind of research needed to catch emerging diseases early. Zhang's lab had the second-highest level of containment, so it could handle most germs that emerged, and he had set up a network of labs across China to routinely monitor viruses.

But when they saw the sequence of the virus from Wuhan two days later, Zhang and his team knew this was not routine but a virus closely related to SARS that had never been seen before. This was the first time emerging disease specialists found out about this: none of the sequences produced for Wuhan hospitals had been publicly posted.

Zhang immediately told the Shanghai health authorities and China's National Health Commission that the new virus was likely to be transmitted via the respiratory tract, as SARS was. This was a big deal: respiratory viruses are especially hard to control.

The National Health Commission triggered a second-level national emergency alert the next day—although it remained a secret from the public. Zhang caught the train to Wuhan, where he saw that the outbreak "was getting serious," he told *Nature* a year later.[17]

Now he faced a dilemma. On January 3, the Chinese government had issued an order requiring all but official state labs to destroy their samples of the virus and release no information about it. Some labs had received the order on the 1st.

On January 7, China told the WHO and then the Chinese public (via state television) that the mystery pneumonia was caused by a coronavirus, but gave no details about it, even though the China CDC later published a sequence it had on January 3.[18]

Zhang sent his sequence to colleagues, but hesitated about publishing. Then on January 11, he was sitting on an airplane when an Australian colleague phoned to ask about posting the sequence. The flight was taxiing for takeoff, and an attendant was asking Zhang to end the call. He thought of what he had seen in Wuhan, he later told *Nature*—and told his colleague to post.

The official response was immediate: later that day, China's National Health Commission finally uploaded its sequences to a public database. The next day Zhang's lab was temporarily shut down by local regulatory officials. It looked like punishment, although Zhang told *Nature* it was part of a normal regulatory process.

The questions through all this, the ones that need answering if we are to understand not just this pandemic but how to manage future ones, are these: What did Beijing know, and when did it know it? Beijing must have known the virus was contagious before January 1, the day China's top health authorities started telling labs to destroy their samples. That was confirmed by what happened on January 3, the day the China CDC had its sequence, when Robert Redfield, head of the US CDC, phoned George Gao, his counterpart in China, to ask about the reports of an outbreak in Wuhan.

Gao, normally an open, unguarded scientist befitting his training at Oxford and Harvard, was reportedly less forthcoming on the call than usual and said Wuhan authorities had told him there was no person-to-person spread.[19] Nonetheless, reported Lawrence Wright of the *New Yorker*, Redfield asked Gao about reports of several cases within families, which are usually the result of person-to-person spread.[20]

We don't know how Gao responded, but immediately after the call, Redfield phoned Alex Azar, the US secretary of health and human services. Azar then told his chief of staff to alert the president's National Security Council that the virus in Wuhan "was a very big deal," the *New York Times* later reported. This is not what you would usually say about a virus that does not spread person to person.[21] "It was clear to Redfield on Jan 3 that transmission was likely," Wright told me.

After that call, Redfield officially asked China if a US CDC team could come and help Wuhan with the investigation. China never issued the invitation. But Wright says Redfield told him that, on a call a few days later, Gao broke down crying and said, "I think we're too late." After that, the calls stopped.

This astonishing revelation suggests that by early January, China not only knew the virus was spreading person to person, but also that it may already have spread too far to control—and the top health authorities in the United States knew, too.

Others guessed: on January 4, the head of the University of Hong Kong's Centre for Infection, Ho Pak-Leung, said that with so many cases in such a short time, the Wuhan outbreak was probably contagious—and Hong Kong started checking travelers for fever.[22] Yet China continued to insist there was no person-to-person spread.

No one could do much about it, though, without a sequence to use to test for the virus. But within twenty-four hours of Zhang's sequence being posted publicly on January 11, Moderna, a vaccine development company in Washington, DC, had designed a test and also started the work that led to its vaccine.[23] Within two days, Thailand had used the sequence to design a highly specific PCR test that diagnosed the first case of the virus outside China, in a woman who had traveled to Thailand from Wuhan.[24]

News that the virus had escaped China changed attitudes in Beijing. On January 14, China's National Health Commission called a teleconference with provincial health officials to warn that the outbreak could become "a major public health event," that the virus might spread explosively during the upcoming New Year's holiday when millions travel, and that "clustered cases suggest human-to-human transmission is possible."[25]

Beijing imposed the top level of public emergency, and health officials were told to test for cases of the virus, open fever clinics, and make medical staff use protective gear. The official case numbers had stood at forty-one since early January, as a major political congress took place in Wuhan; after the meeting on the 14th, they started climbing again.

Yet it was all kept secret from the public, apparently for fear of causing panic. In public, officials still insisted "sustained human to human transmission" was unlikely. The word *sustained* provided cover. Without a serious search for infections in a wide range of people who may have been exposed, only transmission from one known case to another, say within a family, was ever likely to be observed by doctors, and for all they knew such transmission never continued on to more people, becoming "self-sustaining." But wider testing was about to start. Other countries had used the virus's sequence to design what thousands would later know as the PCR test, based on a deep nasal swab. They started screening travelers from Wuhan—and they were finding lots of infected people.

Neil Ferguson and his team at Imperial College London are among the world's most respected mathematical epidemiologists: they construct complex computer models that describe how diseases are observed to behave and then use them to predict how new ones will spread. In January, they used a large database of airline passenger statistics to calculate how many people in the catchment area around Wuhan typically travel internationally.

It stood to reason that the percentage of travelers who were found to be infected should be the same as or less than the percentage of the population back home that was infected, as there was no reason to think people with the virus would be more likely to fly abroad than people without it. But in fact, the percentage of travelers who were infected was much higher than you'd expect based on the official case numbers from Wuhan.

So, they inferred, there must be more infected people in the Wuhan region than was being reported. Imperial crunched the numbers—it's more complicated than simple percentages—and reported on January 17 that there were probably 1,723 cases, give or take, in Wuhan. Wuhan was still officially reporting forty-one.[26]

There was no need to suspect cases were being concealed. The most likely explanation was more straightforward: official numbers counted only people who had a positive test for the virus, and in the early days of the epidemic, the only people being tested were those sick enough to go to a hospital. Other countries, however, were testing every traveler with a fever who had just been in Wuhan, even if they were only mildly ill.

The missing cases might simply have been not serious enough to go to the hospital. They would not, after all, have excited suspicion: mild cases look like flu, and it was flu season.

Still, looking at Ferguson's numbers, that seemed like a lot of cases for a virus that wasn't transmitted person to person, as China was still claiming. Or as the Imperial team drily put it, "Past experience with SARS and MERS-CoV outbreaks of similar scale suggests currently self-sustaining human-to-human transmission should not be ruled out." (MERS, a virus with an even higher kill rate than SARS—around 40 percent—jumped to humans in 2012 and is, like SARS, a close relative of Covid-19.)

More evidence was coming in. Health care workers getting a virus from their patients is good evidence that it spreads between people, and the China CDC reported later that seven doctors and nurses in Wuhan had caught it by January 11. Yet it had sent a team to Wuhan about that time that reported no evidence of human-to-human transmission.[27] On January 10, researchers at the University of Hong Kong found a family over the border in Shenzhen that became infected when they traveled to Wuhan. One family member did not go but became infected after the others came home.[28]

The researchers shared this information with the WHO. On January 15, Japan reported a case in Kanazawa, a patient who had just been in China but had not visited a wet market. The report noted that, according to the WHO, "there are currently cases in which the possibility of limited human-to-human transmission of this disease, including among families, cannot be ruled out. However, there is no clear evidence of sustained human-to-human transmission."[29]

The truth was that China was not testing people and looking for such transmission, or admitting it when evidence emerged. Even though the Hong Kong team reported on January 10 that six people in Shenzhen had tested positive, Shenzhen didn't report one confirmed case, its first, until the 19th.

On January 18, despite growing evidence that the virus was contagious and despite the secret emergency alert, the Wuhan neighborhood of Baibuting staged a potluck dinner with forty thousand people in honor of the kitchen god—and in a bid for a Guinness World Record for the largest number of dishes served at a potluck banquet. The mayor of Wuhan told a television interviewer later, after gatherings of people in Wuhan had been banned, that the party was allowed because they still thought that human-to-human spread was limited.[30] The same day, Wuhan Union Hospital was putting cases in strict isolation.[31]

Then another case turned up in Thailand, again a traveler from Wuhan, who had no connection to Thailand's first case. "Sticking my neck out to the chopping block, I suspect there may already be significant ongoing transmission of this novel coronavirus," Pollack wrote on ProMED. However, most cases were not being reported, possibly because they were mild and unrecognized, Pollack wrote, adding, "I obviously hope I am correct here."[32]

By January 20, cases were being reported across China, Japan, Thailand, and South Korea. Pollack's gloves were off. "It is becoming more difficult to conclude," she wrote testily, "that there has been limited person-to-person transmission as the case numbers are climbing."[33]

Chinese scientists were also losing patience. Also on January 20, Yi Guan, a virologist at the University of Hong Kong who helped uncover the SARS virus, told Caixin that the Wuhan outbreak was behaving like SARS: it was spreading between people.[34]

The same day, China's President Xi Jinping finally went public, telling people to take measures to stop the virus spreading during the coming Lunar New Year holiday. Zhong Nanshan, an epidemiologist called the "hero of SARS" for helping discover the SARS virus in 2003 (and then telling the public it was out of control when Beijing said otherwise), was heading the government's investigation. After Xi spoke, Zhong told China Central Television that the virus spread from person to person.

There were more surprises: the *South China Morning Post* in Hong Kong later reported that, according to classified documents they saw, the earliest recognized case developed symptoms on November 17, not December 1, as later reported.[35] That also fit later genetic analyses of the virus that found it had jumped to humans between mid-October and mid-November 2019.[36] It had taken China a month and a half to spot a problem and tell the WHO. It took another three weeks to tell its own people—and the world—how bad it was.

What happened next shows how out of control things already were in Wuhan by then. To understand that, we have to look at the main ways of fighting an epidemic when you don't have drugs and vaccines: containment and mitigation.

Containment is by far the most effective way to limit an epidemic, if you get to it before there are many cases. The classic method of epidemic control used for centuries is to isolate people with symptoms and then quarantine their contacts for the time it should take for them to incubate the infection and start showing symptoms. Maybe they won't have it—good. But if they do, the quarantine ensures they don't pass it on.

Nowadays, you can, in theory, test people for the pathogen and quarantine only those who test positive—if you really trust your test to produce accurate enough results. In either case, the point is to break the chain of transmission. Do that enough, and you can snuff out a virus: that's how the world defeated SARS.

However, as we quickly learned with Covid, this is a lot harder if the virus can spread before people show symptoms, as neither the person infected nor the people they contact will suspect a problem at the time. And it is hard to do if more than a few people are sick. You have to trace and quarantine all the people each case might have infected, which can add up quickly with a virus as easily transmitted as Covid-19. You won't get everyone, so some new cases will continue to crop up, meaning there will be more people to trace.

It's hard work. As it wrestled the Covid-19 epidemic to a halt in spring 2020, China eventually used six-person teams for each case to track contacts. The European Centre for Disease Prevention and Control estimates it takes a hundred person-hours to track one case's contacts. If you can break all the chains of infection from every case, the disease can be contained.

But you have to start early, before there are too many cases to track. If a disease is spreading generally—in the community—it becomes impossible: not only are there probably too many cases, but people might have no idea whom they caught the virus from. That person could still be out there, spreading the virus, no matter how many known contacts you quarantine of the case you found.

At that point, the standard approach is to switch to what is classically called mitigation. A lot of us know about that now because, with a few notable exceptions, most countries outside China didn't act in time to contain the virus and ended up mitigating: you ban large gatherings, close schools and workplaces, and generally reduce interaction between people to slow the spread of the disease, a set of measures we all now know as social distancing.

At the extreme, as so many of us also know, you lock down and keep people inside. You don't wipe out the virus, but at least it doesn't spread so fast that the sick overwhelm your hospitals. That means the number of cases you get per day does not rise as high or as fast as it would have—the now-famous "flattening the curve."

Early in the Covid-19 epidemic, China discovered that, outside Wuhan and the rest of Hubei province, a mix of mitigation and containment actually worked best: first, case finding and isolation with widespread testing and fever clinics, then tracing and quarantining their contacts to break chains of infection—and then on top of that, if necessary, varying levels of mitigation to slow the spread of the virus, which, because people with the virus had fewer contacts, also made containment more feasible.

But on January 22, Wuhan was already at the point where only lockdown would work. As long as the official story was that the virus did not spread from person to person, officials could not make any visible efforts to limit the spread of infection, back when it might have been possible to contain the virus. Now it wasn't. There were just too many people carrying it.

As a result, China imposed a *cordon sanitaire* around Wuhan, a term from prevaccine days meaning "health barrier." They were invented for cities with the plague so no one would enter—or escape—carrying the disease. (English uses the French term because, in 1821, France revived the concept by sending thirty thousand troops to seal the Spanish border to keep out the yellow fever raging in Barcelona.) No one could enter or leave Wuhan, a city of eleven million, without special permission, beginning at 10:00 a.m. local time on January 23.

That was extended to all of Hubei province a day later. Transport within the city was shut down.

But there was a huge problem: Lunar New Year was only three

days away. This is China's biggest yearly celebration, when 400 million people travel to family celebrations all over the country—the biggest yearly human migration on earth. Moreover, Wuhan is a hub for travel within China. Mass travel had already begun, and at news of the impending shutdown, people flooded into train stations and airports.

Authorities later announced that five million people had left the city before the *cordon sanitaire* could be enforced.[37] Chris Dye of the University of Oxford and colleagues confirmed, using geographically coded mobile phone data, that 4.3 million people left Wuhan between January 11 and the start of the travel ban on the 23rd.[38]

Many were carrying the virus. There was no way to call it back.

Back in Europe, my visitors had gone home, and I was now visiting family in London, with plans to hit the city's January sales. Those plans were dropped when I heard the confirmation that the virus was spreading person to person: I borrowed a desk and emailed my editor and as many scientists as I could. My first report for *New Scientist*, filed January 29, started with the words, "The new coronavirus may be about to go global."[39]

That's how far things had already gone, and it wasn't speculation. Gabriel Leung at Hong Kong University is a leading expert in public health and a veteran of SARS. He and his team had also used phone data to calculate that dozens of infected people had long since traveled from Wuhan to China's bustling metropolises: Beijing, Shanghai, Chongqing, Guangzhou, Shenzhen.

On January 27, he told a press conference that, according to his mathematical models, without "substantial, draconian measures limiting population mobility"—even more restrictive than China had already imposed—epidemics outside China were inevitable. His model forecast two hundred thousand cases by the following week.[40]

Three days earlier, Chinese scientists had published clinical details of the first forty-one patients in the leading British medical journal, *The Lancet*. Chinese doctors complained that the information should have been shared with them earlier, as they started to encounter cases. But it clearly could not have been published while the official story was still that this was nothing like SARS.

"Clinical presentations greatly resemble SARS-CoV. The number of deaths is rising quickly," they wrote. "We are concerned that the novel coronavirus could have acquired the ability for efficient human transmission"—in other words, that it was better than the clumsy SARS virus and had already spread widely. Scientists are good at understatement, but that deserves a prize: the day after the paper appeared, there were officially two thousand tested and confirmed cases across China and, we can now calculate, at least eight thousand milder ones, probably far more.

The Chinese scientists were clear about what was needed to manage the epidemic: reliable, quick tests for the virus. They also noted the discovery, in 2013, by the Wuhan Institute of Virology, of very similar viruses in bats that were already capable of infecting human airway cells.

"Because of the pandemic potential of 2019-nCoV," they warned, using the original name scientists gave the virus, it would have to be watched carefully to see how its transmission and impact changed as it adapted to humans.[41] It could be expected to evolve, and as we now know, it did.

Everything was there. Efficient spread. Need for tests. Pandemic potential. At that point, countries around the world should have started intensively preparing for the virus to hit. Some did. Most did not.

So what we can conclude is that despite its apparent openness, China delayed reporting the illness, the virus, and especially the all-important person-to-person spread for weeks, even though top officials had that information. They had the sequence in late December and admitted the severity of the problem among themselves on January 14, but it was six more days before they told the public—and people were traveling out of Wuhan the whole time. From comments made by officials about social stability, it seems likely that they were simply afraid of public panic—or perhaps of looking bad to superiors.

Those fears had tragic global consequences, but what is even more frightening is that they certainly weren't the first government officials to downplay an outbreak of infectious disease for those reasons. Nor is the phenomenon unique to China.

In any case, the mayor of Wuhan ultimately had to resign and accept responsibility for missteps, although before he did, he blamed Beijing for controlling what he could say about the virus in public. Those controls don't seem entirely gone. Ai Fen's March 2020 interview in *Renwu* reportedly keeps mysteriously disappearing from Chinese websites. It has been kept alive by Western coverage and Chinese netizens.

Meanwhile, the virus Ai was told to shut up about has gone everywhere. On March 11, WHO director-general Tedros Ghebreyesus declared it a pandemic.

So that, as far as I can piece together at this time from a range of reports, is what happened. It is of course probable that more will come to light, and accounts could change. But now we can start to ask the crucial questions: Could this all have been prevented? Could the Wuhan outbreak have been stopped from spawning a pandemic?

This is one of the first big outbreaks that has been analyzed as it happened, using modern technology for rapidly sequencing viruses from different patients and working out which virus is descended from which on the basis of small, shared mutations. And what stands out is that the first few published sequences from patients in China, says Andrew Rambaut of the University of Edinburgh—who specializes in the evolution of emerging RNA viruses like this—were "genetically very similar."

The longer a virus circulates in a species, the more it acquires random small changes in the gene sequence. If this virus had jumped several times from different animals, or circulated for a long time in people before it was spotted, there would have been more genetic variation in the early infections.

So, says Rambaut, "I would say that it was definitely a single jump." It could have been from one animal to one human, or it could have been from a few animals carrying the same virus to a few humans—without details of the earliest cases, he says, we can't tell. A WHO mission was shown further examples of genetic variation in early cases in March 2021, but Rambaut says the data still suggest a single source, followed by human-to-human spread.

That means there was no hidden epidemic happening over a much larger area or longer time, or there would have been more genetic variation by the time of the earliest sequenced viruses. That means those first cases in and around Wuhan were all there were. In theory, if Wuhan had enacted stringent containment the moment they spotted an unusual cluster of pneumonia cases, in late December, then actively looked for other infections and contained them while there weren't many, they might have stopped the infection from spreading far. It would have been even better if they had spotted it earlier. To answer the question of whether they could have stopped the virus from spreading entirely, however, we need to know how much action would have been needed and whether authorities would have agreed to the disruption it would have caused, knowing what they knew at the time.

Andy Tatem and Shengjie Lai and their team at Southampton University in England have measured how authorities might have done it. Cases in China increased exponentially, as diseases do when there is nothing to stop them, until the *cordon sanitaire* was thrown around Wuhan. After that, and as similar travel bans and social distancing orders were imposed across China, case numbers stopped rising.

The impact was stunning: China's epidemic actually peaked in mid-February, a turning point predicted by epidemiologists outside China on the basis of changes in reported case numbers as controls were imposed, and confirmed by a WHO delegation to China in late February. By late March, China was reporting no new cases. The problem was now everywhere else.

In a dizzying analysis using a mathematical epidemic model and the seven billion anonymous location records per day logged by China's Baidu cell phone network, Tatem's team quantified how people moved between China's 340 major cities as travel restrictions went into effect after January 23. They measured how that travel related to the data on the spread of the virus. Given that, they then worked out how the virus would have spread if travel was the same as Baidu had logged during the same weeks in previous, normal years, with no travel bans.

They calculated that with unimpeded travel, provinces outside Hubei would have had *125 times* more cases by the end of February. "China's vigorous, multi-faceted response is likely to have prevented a far worse situation, which would have accelerated spread globally," they wrote.[42] There would have been much more of the virus in the world at that early stage—what epidemiologists call amplification—if China had not wrestled its epidemic to a halt when it did.

And if it had started containment measures only a week or two later, it might not have been able to do that. We have now all seen how fast an exponential epidemic shoots up. If Zhang Yongzhen had not posted his genetic sequence when he did, Thailand would not have been able to detect its first case two days later, rattling Beijing into action. Any later, and the situation in China might have been beyond containment.

But if Wuhan had imposed travel bans before those five million people left Wuhan for the Lunar New Year holiday, could it have stopped the virus entirely? Tatem's team found that if China had imposed the same control measures a week before January

23—just after the crisis meeting over the Thai case—it would have prevented 67 percent of its epidemic.

And implementing the control measures from early January— when Chinese authorities knew the virus was contagious and had the gene sequence to base tests on—would have cut China's epidemic to only 5 percent as many infections. Such a small epidemic might well have been contained, especially if other countries were also alerted to watch for, test, and contain any infected people who turned up crossing—or within—their borders.

"Technically we certainly could have attacked it effectively at that point and maybe contained it," says Tatem. "It's easy to say this with hindsight of course. There was so little we knew about the virus at that point. That would have made it hard to act rapidly."

Rambaut thinks more could have been done. "The authorities in Wuhan spotted the outbreak as an unusual cluster of pneumonia," he says, "but then spent weeks saying there was no evidence of human-to-human transmission when in fact it was doing exactly that." As it was, China acted barely in time to contain its own epidemic, and well after the virus had spread outside its borders.[43]

An independent panel reporting to the WHO concluded rather drily in January 2021 that "public health measures could have been applied more forcefully by local and national health authorities in China in January."[44] All that would have been needed, says Rambaut, would have been surveillance to spot the outbreak early, then intensive containment and contact tracing to break all the chains of transmission before there were many cases.

In fact, those were tools China already had. In 2003, the SARS coronavirus initially spread out of control in China and eventually

29

to countries around the world, because doctors' initial warnings about the outbreak were stifled, at first simply by local bureaucratic inertia. To stop that from ever happening again, in 2004, China installed the National Notifiable Disease Reporting System in every hospital.

Doctors were required to enter the diagnosis into the system whenever they encountered certain key infectious diseases, including pneumonia of unknown origin, the *New York Times* reported on March 29. A suspicious cluster would show up on a screen at the China CDC in Beijing without anyone having to get past any reluctant local bureaucrats.[45]

If something worrying appeared, central officials could launch intensive case finding and containment efforts. In an online drill in July 2019, 8,200 health officials tracked and contained a simulated infection brought in by a traveler who had been registered on the system.

There was a compelling reason for doing this besides avoiding a rerun of SARS. Several strains of bird flu that can infect and kill people have emerged in China over the past twenty-five years—we'll be looking at those later. The one saving grace of these bird flu viruses so far is that they cannot transmit between people, although research has shown they can evolve that ability. If one became transmissible, it might be catastrophic. A cluster of cases suggesting the emergence of a transmissible strain would have to be contained with extreme urgency.

Reflecting that, doctors were instructed to enter any case of bird flu they encountered in the national direct reporting system within two hours of diagnosis. The frequency with which individual bird flu cases have been diagnosed across China over the past decade—to judge from ProMED—suggests the system has been working. Luckily, no worrying cluster has yet emerged.

Perhaps when tests showed that the unusual pneumonia cases in Wuhan in November and December 2019 were not a new kind of flu, health officials relaxed. According to leaked internal reports, in December 2019, doctors were told not to report such cases to the automated alert system, only to local health officials, who were reluctant to pass on bad news. They were more reluctant during the local Party Congress in Wuhan that January, during which the official case numbers remained fixed at forty-one. Chinese officials effectively confirmed this when they told the Joint WHO-China Study Team in early 2021 that Covid was not made reportable to the system until January 20—whereupon it got 174 reports, just of cases that started in December.[46]

It was as if someone took the batteries out of the smoke alarm that sounded too many false alarms—so it missed a real fire. Word may not have reached Beijing of the mystery pneumonia until late December. But that didn't trigger a big effort to find and contain cases. According to reports in the Chinese press cited by the *New York Times*, officials set a case definition from January 3 that was virtually designed to minimize detection of a spreading infection: doctors could test cases of pneumonia for the new virus only if the patient had had some connection to the now-closed wet market or to a known patient with a market connection—a strange case definition for a virus that isn't supposed to spread person to person. In Wuhan, the virus was spreading freely, so, increasingly, people who got it did not have the requisite connections to be tested. The

WHO China team was told the requirement for a market connection was only dropped on January 18.

That could be why, until then, Wuhan's official case numbers stopped climbing: people had the disease but weren't being tested. It's worth pointing out that this happened elsewhere too: later, some US states and European countries would test people with Covid-19 symptoms only if they had contact with China or a known case, well after the virus was already circulating elsewhere—including locally. As a result, they turned out later to have far more cases than they had realized.

But could the early warning system have triggered enough action, early enough, to contain the disease? The system was designed to trigger a full-blown containment response. The cases in December should have been enough to do that.

Would local officials have followed through with something that was not a drill? This raises a perennial dilemma of public health, as I heard from Sylvie Briand, head of infectious hazard management at the WHO, when we were discussing problems like this some months before she was plunged into the Covid-19 crisis. Containing a new infectious disease before it has spread far nearly always means reacting before it seems like a big deal, she says. There may be only a few clinical cases, but you know several times that number are already infected and incubating the disease, especially if it is very contagious and spreads early in the course of infection. Covid-19 ticks both boxes. You have to contain such things early before they escalate.

This can be difficult, as at that point officials often see the threat as too trivial for such disruption, scoffing that more people die falling down stairs—people actually said that in the early days of Covid-19.[47] The slight problem with that analogy is that stair accidents don't then multiply exponentially. But if you do manage to act early and contain the disease, nothing happens. Officials may wonder why they spent all that money fighting a threat that disappeared, even though that was the point. I still get letters when I write about a new disease, saying something like, "Oh, well, SARS was supposed to kill us all, and it never did, so why should we believe this?" Well, because with SARS, we eventually listened to the warnings and managed to contain it. We were also lucky.

Wait until the threat is obvious, though, and you are usually too late. "First they accuse you of overreacting," said Briand, echoing many frustrated public health experts I have heard over the years. "Then the epidemic suddenly explodes, and they say you didn't act fast enough." Research has actually shown that few people understand how fast exponential really is, and that when the exponential spread of the virus was explained early in the pandemic, support for Covid control measures increased.[48,49]

In fact, the problem of getting official backing for early controls, then blame if nothing bad happens, may explain officials' repeated reluctance to tell people the truth about outbreaks "for fear of causing panic." It isn't really about panic, and it isn't confined to China. Veteran journalist Bob Woodward reports that in February 2020, when the United States had only thirteen known cases, Tony Fauci told a closed governors' meeting that Covid posed a "very serious" risk and later spoke of having scared them. Yet the press release on the meeting said the risk to the US public was low, and two days later Fauci said the same thing to a public conference. When challenged, Woodward writes, Fauci agreed Covid might go pandemic, but if he proposed extreme remedies too soon, he would lose credibility, preventing subsequent action.[50] "It was surreal," recalls Helen Branswell of Stat News. "The *Titanic* had a gaping hole, but we were drinking tea on the deck."[51] This communications problem is something we need to fix to make governments respond in time to future outbreaks.

But getting that support is especially hard when you cannot contain a virus just by quietly isolating the few people who have it and the few dozen people they contacted closely enough to pass it on. Officials and the public might be okay with something that minor. But it might not be so easy.

Using a massive set of data on real social interactions in the UK, Matt Keeling at the University of Warwick and colleagues found that using the official UK definition of contact—being within two meters of someone for at least fifteen minutes—you have to trace and quarantine thirty-six people per case of Covid-19 just to catch and isolate four of every five people that case infected.[52] That's a lot.

And contact tracing might not be enough. As we have seen, the Chinese subsequently discovered that the key to stopping Covid-19 is using social distancing as well as containment. The variable that matters—and the one piece of epidemiology jargon you really need to know to get all this—is R0, "R nought," the basic reproductive number. It was a previously arcane term many nonepidemiologists got familiar with in 2020.

This is the number of people each infected person passes the virus on to, on average, at the start, when everyone is susceptible, and no measures are yet being taken to slow contagion. And we all were susceptible to start with, as this was a virus no one had encountered before.

That value for Covid-19 was originally calculated at between 2 and 3, making it more communicable than most seasonal flu, although, as some experts predicted from the start, some people or situations might cause much more transmission than that, in what we all learned to call superspreading events.[53] Even without those, Rosalind Eggo at the London School of Hygiene and Tropical Medicine and her team calculated that for a virus with a basic R0 like the one that causes Covid, contact tracing and isolation alone work only if there is little or no transmission before the virus causes symptoms.[54]

Otherwise, an infected person will have too many untraceable contacts because the contact occurred before they knew they were sick. And even if you find those contacts, they will have had more time for their own infections to incubate and may already have spread it before you can quarantine them. We now know that around half of all people with Covid caught the virus from people who were not showing symptoms at the time. Without very wide-spread testing, that poses problems for containment.

It's as if a virus with a high enough R0 is just too slippery to pin down easily. So the answer is to slash transmission by reducing the number of people contacted by each infected person. This doesn't change the virus's intrinsic R0, but it changes its effective transmission, usually just called its R value. The Chinese discovered early that social distancing does this, making containment feasible: with less person-to-person contact, fewer people are exposed to any given case, so you have to quarantine fewer people to break transmission.

If a virus has an R0 around 2.5, Eggo and team figure you need to reduce contacts by around 60 percent to get the R value down to 1, the level at which the epidemic stops growing. With more distancing, cases drop. But the moment you relax distancing, the R value and cases rebound. The repeated mistake of Covid management has been using lockdown to get case numbers down to where containment becomes possible, then not following up with adequate containment to keep them down. Time and time again, case numbers rebounded, wasting the personal, social, and economic sacrifices of lockdown.

This also means that, even if Wuhan had thrown itself into case finding, contact tracing, and quarantining at the very start, it might well have been insufficient to stop the epidemic, without some social distancing as well to make containment easier. That is certainly what other Chinese cities discovered later. But in early January in Wuhan, epidemiologists would probably have known too little about the virus to have realized the need, or at least to make a political case for such drastic measures. Even much later, with far less excuse, some Western countries were slow to admit the need for such disruption—to put it very politely.

"Social distancing is the magic ingredient in control," says epidemiologist David Fisman of the University of Toronto, another SARS veteran. "I have no reason to think they could reasonably have known that massive social distancing was needed in response to what at first appeared to be just a disease cluster in Wuhan."

That's the trouble with a new disease, he says. "We all learn a lot week by week and are all making mistakes. That's the nature of the beast, I think."

Tatem agrees: "You only have to look back at ProMED to see lots of unexplained small outbreaks that lead nowhere," like those earlier reports of undiagnosed pneumonia from China. In fact Covid's dynamics suggest that more than two-thirds of similar events, with the virus jumping to humans, would have died out before getting far, like trying to make a spark catch in damp kindling.[55] With Covid, we got unlucky—it eventually caught. But we can't throw a city into lockdown for every new virus that turns up in surveillance. How do we distinguish the real threats so that we risk mass disruption only—or at least mostly—for things that aren't going to fizzle out?

"We do need to get better at early detection and identification of those outliers that have the potential to cause major outbreaks," says Tatem. But of course we can't even try to decide which outbreaks really have legs if we don't know the outbreaks are happening in the first place. That's where China's smoke alarm should have worked.

Zeng Guang, the chief epidemiologist at the China CDC, is quoted as telling the Communist Party paper *Global Times* in early 2020 that local governments in China "only partially" based their decisions on what the scientists told them, instead favoring "social stability, the economy and whether people could happily enjoy Lunar New Year."[56] You can't cause much disruption if that's your aim.

A preference for secrecy and stability won out over the scientists' epidemic models at the typical crisis point in public health: when you need strong action, even though it doesn't look to observers—or politicians facing the year's biggest holiday—like much is wrong.

So, the big question: Could China have stopped the epidemic from becoming the pandemic? The epidemiology suggests that they might well have been able to slow it down, although it might have been hard to stop completely, even if the automated system had been allowed to do its job. Just the effort, though, would have had an incalculable impact.

It would have meant telling the world that a dangerous, contagious pneumonia had emerged in Wuhan. If, over at ProMED, Marjorie Pollack had been able to post that in December or even January 1, and the WHO had announced it, the world's virologists and epidemiologists would have run for their labs and models and started furiously posting results, as indeed they did a few weeks later once the news was out.

The world's developers of vaccines and drugs and especially diagnostic tests would have got to work. Other countries could have started earlier at testing people who had traveled to Wuhan—plus the people who might have contacted them, or anyone who turned up with Covid symptoms. As more cases appeared, China might have seen the need for social distancing and containment, perhaps before five million people carried the virus out of Wuhan.

Those things happened eventually, but an earlier warning would have given everyone a few weeks' head start. We've all seen now what exponential looks like. A short time, at the right time, matters.

There is no question that when China finally did act, it was awesomely effective, if socially and economically painful. Dye's team found that, normally, 6.7 million people travel out of Wuhan in the month after New Year's. In 2020 there was almost no movement. That bought other cities, and the world, time to prepare.

Eventually, 136 Chinese cities shut down their public transport and 220 banned mass gatherings. Dye's team found that cities that did those things sooner rather than later had a third fewer cases during the first week of their outbreak: curves were flattened, and the number of cases each person infected, the R value, was slashed. Their models showed that the Wuhan travel ban alone or the shutdowns in other cities alone would not have reversed the epidemic curve that was rising fast in China when the alarm was finally sounded, but both together did—and cut the cases China would otherwise have had by 96 percent.

It has not been widely appreciated outside China, however, that, as the WHO later reported, only Wuhan and its neighboring cities in Hubei had to completely lock down. In other provinces, cities applied varying levels of distancing measures. Stores took people's temperatures before allowing them in. Anyone with a fever could go to a "fever clinic" for testing. One of the key measures was that people with cases too mild for hospitalization but no room to isolate at home—the majority—were isolated in repurposed hotels, stadiums, and conference centers, known as Fangcang hospitals.[57] Contacts of infected people were traced and quarantined. An international team led by the WHO went to study China's response to the epidemic in late February. They reported that China had successfully managed to bend a steadily rising epidemic curve sharply downward. It stopped the virus spreading in the community in every province outside Hubei—most transmission was within families. It was, by any standards, an amazing achievement.[58]

Bruce Aylward, the Canadian epidemiologist who led the WHO

team, was jet-lagged enough to betray more than the usual slight hint of his Newfoundland accent when he briefed the press the day he flew back from Beijing. But he said he was convinced the decline in case numbers was real. Doctors had spoken of scheduling ordinary patients again. Lines outside fever clinics had disappeared. A big trial testing an existing antiviral drug against Covid-19 was having trouble finding participants.

China's initial delay may have let the virus get away. But its subsequent massive crackdown bought the world time, said Aylward. If the spread of Covid-19 out of China was terrifyingly fast, don't even try to imagine what it would have been without the anchors China slammed on its own epidemic.

But most importantly, "we now know what works against this virus. We know what to do," Aylward said. He dismissed claims that only China could have imposed the containment and social distancing needed—the rest of the world could follow their model, adapting measures to their own conditions. Events proved him right, as numerous countries eventually imposed lockdowns as strict or stricter than much of China ever needed. But Aylward was also right about another thing: he wasn't sure the rest of the world "understands the need for speed."

Most of it did not. The virus already had a head start in Italy, the UK, the United States, and elsewhere by the time those countries launched a serious response, and case numbers there rapidly pulled ahead of China. In late March, no Chinese province outside Hubei had officially reported more than 1,500 confirmed cases, but fifteen US states had—and most Chinese provinces have more people. And tragically, US case numbers just kept climbing.

Some places, however, contained the virus effectively, many without the disruptive lockdowns needed in Hubei and the West. We all know the honor roll: Hong Kong, South Korea, Singapore, Taiwan, New Zealand, Vietnam, Ghana, and a few more. All imposed containment early enough to have a chance of working and backed it with social distancing as needed and widespread testing. Their success suggests what might have happened in China if it had let its disease reporting system trigger a massive containment effort in response to the first case cluster.

Those countries also leveled with people. In an astonishing public statement, Prime Minister Lee Hsien Loong told Singaporeans on Facebook as early as February 8 that, despite a strong containment effort, the virus would probably spread in the community, and he outlined the self-isolation measures that would be needed, "so we will be mentally prepared."[59] In contrast, when Nancy Messonier of the CDC in the United States issued a similar warning on February 25—which rattled stock markets—she was removed from public view.[60]

Singapore pulled no punches, however. "Fear can make us . . . do things that make matters worse, like hoarding facemasks or food, or blaming particular groups for the outbreak," said Lee. On the other hand, he described efforts already happening in Singapore to help people in quarantine, while businesses, unions, and public transport were "going the extra mile" to keep things running. "This is who we are," he declared.[61] At a time when some countries were in denial about the virus, it was a moving performance. But it wasn't just emotional. Leadership mattered. Public trust, say experts in disaster management, is essential for responding to a crisis. And for that, the government has to trust people with the truth. That seems hard for some governments, especially when it concerns disease.

Some, however, managed. On March 22, 2020, the president of Ghana spelled out what the country had to do to control infection, including targeted lockdowns and army and police assistance with contact tracing. He famously explained that, while this would cause hardship, "we know what to do to bring our economy back to life. What we do not know how to do is to bring people back to life."[62] Social scientists studying countries' pandemic response found that when leaders provided clear, early, expert-led and empathetic guidance, their countries had lower death rates than countries where leaders politicized the virus and lost public confidence with abrupt policy changes and overstated optimism.[63] By March 2021, Singapore and Ghana had had 5 and 22 deaths per million of their population, respectively. Those numbers for Brazil, the United States, and the UK were, respectively, 1,272, 1,600, and 1,863.[64]

While leadership was crucial, other factors helped. The key for some of the successful countries was experience with a similar disease. In 2015, South Korea had an outbreak of MERS, which they got under control using hospital infection control and quarantine. And Hong Kong, South Korea, Singapore, and Taiwan were all hit hard by SARS. They knew the need for speed.

Hong Kong traced and quarantined contacts, closed schools, canceled big events, quarantined arrivals from affected countries, and encouraged working from home—all actions other places resisted but had to do eventually. At the end of March, it had only 715 confirmed cases and 4 deaths. The measures actually cut the transmission of flu at the same time by nearly half.[65] Ordinary people's behavior—masks and social distancing—made the difference, something David Fisman says is true in most epidemics.

At university lectures in Singapore in March, a maximum of fifty students were allowed, they sat at two-meter distances, and a photo was taken of who sat where in case contacts needed to be traced later—measures that later became commonplace elsewhere but seemed astonishing at the time. Public spaces were not closed, but anyone entering one had their temperature taken, increasing public confidence as much as catching cases.

South Korea had companies making Covid-19 tests by early February. National labs double-checked the test results as people were tested, effectively doing the usual validation trials of a new test on the hoof to save time. The US Food and Drug Administration insisted that trials be done on US-made tests before they were used on the public, adding to an already disastrous delay in testing.

On top of that, South Korea had invented drive-through testing by late March. Positives were isolated and contacts quarantined, and unlike some places that later used those measures, South Korea enforced them. By April, case numbers were falling, without severe social distancing. The story was similar in Singapore and Taiwan. The difference was the early start that China had missed. Digital privacy experts have valid concerns about the expanded electronic surveillance involved, and studying the options for managing contact tracing and isolation electronically without undue invasion of privacy is now a growth industry. But the virus was contained.

You didn't need a history of tangling with coronaviruses to do the right thing, though. The small Italian town of Vò in Lombardy kept the virus under control by testing everyone and then imposing isolation and quarantine as needed.[66] This should have been possible in far more countries in that first onslaught of contagion, yet many utterly failed.

If nothing else, these successful responses demonstrated that containment, started early enough, worked on Covid-19. They confirmed that earlier action in China might have limited the epidemic. But mistakes were far from unique to China.

Wuhan had its Guinness-record potluck. But on March 7, as the pandemic took hold in France—and we all knew the virus was contagious—more than thirty-five thousand people dressed as Smurfs gathered in Landerneau. The next day, France banned gatherings of more than one thousand people.

In late March 2020, seventy University of Texas students were among hundreds who crowded onto beaches for the traditional spring break, despite warnings; forty-four of the seventy later tested positive for Covid-19 and undoubtedly gave it to others. All these reactions seem to be simple, psychological denial: the refusal of people who have rarely been at much risk from infectious disease to believe they really need to take a so-far largely invisible threat seriously.

Five million people left Wuhan before the lockdown. But even that painful lesson was not learned in time to avoid it elsewhere. More than six weeks later, Italian authorities locked down northern provinces that were the initial hotspots for the virus. The news leaked the evening before, and people fled, carrying the virus all over Italy. The whole country was shut down the next day. It seemed unthinkable at the time. But it was still too late to prevent Italy's otherwise enviable health system from being tragically overwhelmed.

In many countries, social distancing was partial or delayed, to the point where curves were barely flattened. Testing was delayed or restricted, endangering health care workers and patients and preventing containment. Even as the WHO stressed that containment worked with this virus, some countries abandoned it almost immediately, including Switzerland, where the WHO is located.

And ideology trumped public health in many places. A US administration focused on threats from foreigners rushed to close borders, then did little more—although the virus was already spreading within the United States.

My purpose in writing this book is not to analyze what countries did beyond the very early days to respond to Covid-19. Those much-needed analyses have been going on in media, scholarly journals, national governments, and neighborhood Zoom meetings since the pandemic took hold. For now, we can say that few covered themselves in glory, and the steep learning curve for pandemic control in many places came at a high cost in death and chronic illness. The accusations and political fallout will rage for a generation.

For now, we can ask whether more openness and earlier containment in China might have prevented the pandemic. This is not to point fingers or throw stones—if the pandemic taught us anything, it is that most of us, in that respect, live in glass houses—but so that the next time this happens, wherever it happens, we might do better.

The answer seems to be that stopping Covid-19 entirely might have taken faster action than any government could have managed. But earlier action was possible, and that might have slowed the epidemic enough to make Covid-19 much less damaging and perhaps, just perhaps, kept it from reaching pandemic proportions.

According to the Chinese Communist Party's official newspaper,

China's Supreme Court admitted as much on January 29 when it ruled that authorities in Wuhan were wrong to censure the eight doctors for their online chat about a SARS-like virus back in December. "The information would have pushed the public to take preventive measures more promptly, which could have been a fortunate thing given the current efforts needed to contain the virus."[67] Xi's government even turned Li Wenliang into a posthumous hero.

The first official case in Italy was detected on February 20. Italian public health officials did the right things: isolating, contact tracing, locking down towns with the most cases. But it was too late: the virus was already too widespread, and hospitals were eventually overwhelmed. In fact, Italian epidemiologists later discovered that the first traceable case in the country fell ill on the first of January,[68] while analysis of sewage samples suggests the virus arrived even earlier.[69] At the time, no one suspected a thing.

If every country had known what China knew by early January, if it had sounded the alarm openly and much sooner and shared everything it knew, what might we all have done to stop the virus?

We will be looking at ways all of us might try to do better next time. Through pandemic planning. Through global viral surveillance and response when we find something worrying. Through a binding international agreement to monitor and control pathogens, this time with teeth. Through scaring the daylights out of ourselves by looking at what a worse pandemic might do.

First, let me explain why I'm so sure this will happen again. Let's look at where these viruses come from.

# CHAPTER 2

# What Are These Emerging Diseases, and Why Are They Emerging?

> A new disease every day, and the old ones
> are coming back.
>
> —Loudon Wainwright III, "Hard Day
> on the Planet"

Ever since the HIV pandemic, people from health experts to screenwriters have been predicting what the next one would be. Various kinds of flu? A super-transmissible Ebola with wings? A souped-up version of the common cold? A bioweapon or a therapeutic virus gone wrong?

There have been scares and near-misses with bird flu, mad cow disease, Ebola, SARS, and MERS, then the swine flu pandemic in 2009 that turned out relatively mild, although it still killed. Now there's Covid-19.

Why is this happening, and will it continue? And more to the point, what will hit us next? There are more viruses where Covid-19 came from.

We should start by saying what we mean by pandemic and disease events generally. An outbreak is one or a few cases of an infectious disease that is unusual, so it gets noticed. An epidemic is a larger version of that: more related cases of a disease than usual, spreading in a group of people. An epidemic can be a surprise, or a regular event—for example, the spread of flu through a city in the winter. Endemic means a disease that rumbles on all the time, like tuberculosis or gonorrhea.

A pandemic is when an epidemic goes global. Some health authorities impose other criteria, such as that it has to be severe, or uncontrollable, or new, but those are not consistent or universal. In fact, there aren't fixed criteria for when an epidemic is big enough to count as a pandemic, except for flu—and even those have been changed lately.

We are sure there will be more pandemics, though, however we define them, and probably sooner rather than later, as our population grows and we increasingly crowd into cities, and as intensive global trade and travel continue to transport the illnesses that emerge to people everywhere—especially if we continue to do all that without routine surveillance of the world's scarier viruses or agreed global means for working together to contain them. The chaotic global response to Covid-19—just the fact that it got away in the first place—should make that situation obvious.

We can't precisely predict which disease will be next, or when, although anyone who studies infectious diseases could, and in fact did, tell you years ago that coronaviruses were prime suspects.

There's even an official short list of the diseases we're worried about but aren't ready for. To understand this predicament, though, we need to look back at some recent history.

In 1972, one of the world's then-leading experts on human infections, Macfarlane Burnet, co-wrote the fourth edition of a medical textbook called *Natural History of Infectious Disease*. And in it he wrote something quite astonishing. "The most likely forecast about the future of infectious diseases," he opined, "is that it will be very dull."[1]

He must have enjoyed the shock value of that statement. Scientists always talk up their subject area, especially when trying to entice students into it. Burnet had just won a Nobel Prize for helping work out how our immune systems attack germs, but not us. He knew the value of studying infectious disease.

But the point of studying infectious disease is so you can beat it. And he figured it had been beaten. The comment was an inside joke, a somewhat smug whoop of victory. And also advice to a young doctor: specialize in something else. We've got this.

After all, in 1972, smallpox was all but eradicated. Most childhood diseases, even the super-infectious measles and the dreaded polio, could be entirely prevented with vaccines and had been in richer countries. Formerly deadly bacterial foes—diphtheria, anthrax, TB, typhus, syphilis, gonorrhea—could be killed with antibiotics. Cheap, easily obtainable drugs stopped you from getting malaria. Infections still plagued poor countries, but surely development would fix that. In the 1970s, the Yale and Harvard medical schools downsized their infectious disease departments.

Burnet acknowledged, of course—because he knew colleagues

would cavil if he didn't—that there was always a risk of "some wholly unexpected emergence of a new and dangerous infectious disease." But this struck him as unlikely. "Nothing of the sort," he assured the reader, "has marked the last fifty years."

How about the subsequent nearly fifty years, then? Let's think. The deadly bacteria of Legionnaires' disease appeared four years after he wrote that. The United States recognized the AIDS pandemic four years after that.

Then Lyme disease. SARS. MERS. Ebola. Marburg. Bird flu. Swine flu, another pandemic. Dengue. Chikungunya. Zika. Sin Nombre hantavirus. Nipah. Hendra. Lethal versions of normally harmless *E. coli* bacteria. Gonorrhea that resists all antibiotics. Ordinary urinary tract infections that resist all antibiotics. Extensively drug-resistant TB. West Nile. Mad cow disease, in cows and people. Oh yes, and a Covid-19 pandemic.

I wonder what Burnet would have made of the year 2020. Covid-19 is a lot of things, but it isn't dull.

What an eminent scientist thought in 1972 might seem like ancient history, but it matters to the Covid-19 saga. After infectious disease faded as a leading cause of death, and people everywhere started living to previously rare old age, the big killers—in rich and increasingly in poor countries—became conditions linked not to pathogens but to genes, environment, and lifestyle: cancer, heart attacks, strokes, Alzheimer's, traffic accidents, the complications of smoking and obesity. (There is very recent evidence that bacteria may be involved in big killers like Alzheimer's and heart attacks, but that's a different matter.[2])

Tackling those challenges didn't require the community-level public health that was historically designed for communicable disease, which involved quarantines and vaccination drives, not admonitions to eat more vegetables. The new killers certainly didn't require investment in new vaccines or antimicrobial drugs, pathogen surveillance, or local agencies that could monitor and contain epidemics. Nearly all of those capabilities atrophied, even in the richest countries.

Despite increasing alarm among researchers and global health experts about emerging infectious disease for almost three decades now, the mainstream attitude, especially in rich countries, has been complacency—perhaps because, as always in public health, problems tend to be invisible until it's too late, unless you're an expert and looking carefully. The old infections seemed to be gone, or problems only for the poor or marginalized. New infections seemed merely theoretical.

In response, medical industry shifted. Vaccines used to be made by government agencies as a public good, not for profit. The vaccine that eradicated smallpox, for example, was largely made by the Soviet Union and New York. By the 1980s, vaccines were privatized, and in many cases, profits have been too low to encourage new investment. Most flu vaccines are still made using chicken eggs, a slow and sometimes problematic process from the 1940s.

Investment in public health decreased in many countries. In the United States, there was a brief infusion of cash into preparedness for perceived bioterror threats after the anthrax mailings of 2001. But funding for the Public Health Emergency Preparedness agreement among state and federal agencies fell from nearly a billion dollars in 2002 to $675 million in 2019.[3]

The minimal importance accorded public health was reflected in widespread cuts after the financial crisis of 2008. There has been a surge in hepatitis, Legionnaires', and infections transmitted by sex or drinking water across the United States, which public health experts attribute to health departments losing a fifth of their employees during that time.[4] This hampered efforts to contain Covid-19.

In Europe, too, investment in public health plummeted after 2008. In 2019, a British think tank calculated that public health spending in England had fallen by £870 million just since 2014 and that this may have caused 130,000 deaths and a rise in chronic conditions, like diabetes, that incidentally also make you more likely to die from Covid-19.[5]

The same thing happened with infectious disease monitoring and research in developing countries, where a network of labs largely left over from the colonial era disappeared, seen by former colonial powers as expensive anachronisms in the 1970s. Among those decommissioned was the British lab in Uganda that identified Zika and thirty other new viruses between 1930 and 1970. What if that lab had survived, increasingly with Ugandan scientists in charge, to recognize HIV in the 1970s? As we all know now, early action can make huge differences in a pandemic.

To appreciate the short-sightedness of this dismissal of infectious disease, let's look at some actual ancient history. In the ten millennia since we invented agriculture, infection has been by far the biggest killer of humanity, despite impressive competition from war and famine.

According to virologist Ab Osterhaus of the Research Center for Emerging Infections and Zoonoses in Hannover, in 1900, infectious disease caused fully half of all human deaths.[6] Malaria alone is thought to have killed half of all humans who ever lived. (Those statistics don't contradict: the deaths involved different human populations over different times.)

In the 1800s, tuberculosis, known as TB, infected 70 percent to 90 percent of Europe's city dwellers, and while many of those infections remained quiescent, the bacteria still caused more than a third of all deaths, spawning a score of "consumptive" characters in Victorian novels.[7] Yellow fever killed most of Napoleon's army in the Caribbean, so he unloaded the Louisiana Purchase on the United States and abandoned the unhealthy New World.[8] Many, if not most, children until very recently died before the age of five, nearly always of infection. In some places, they still do.

Both illnesses and deaths by infectious disease in the industrialized world plummeted after 1950 and declined in many developing countries, too. By 2004, infectious disease caused less than a quarter of all deaths worldwide, mostly in poor tropical countries. In rich temperate countries, it was barely a few percent.[9]

Many things contributed to this astounding decline. Besides drugs and vaccines, there were sanitation and hygiene. There was also vastly better nutrition, as chemical fertilizers and crop breeding boosted agricultural yields, and refrigeration and railways distributed fresh food, with the added bonus of banishing disease-ridden livestock, like milk cows with TB, from cities.

That's the big picture. There are countless little ones. I'm just old enough to have had measles at three, and I'm told I almost died of the common bacterial complications, saved by a large—and, I clearly recall, painful—midnight injection of what my mother said was penicillin. A few years later, my little brother got the new measles vaccine.

Mothers who listen to the lies of anti-vax campaigners today have never seen how measles, typhoid, and polio can carry off children. Afghan mothers have: in 2006, as aid agencies tried to remedy years of atrocious health care under the Taliban-led government, mothers and children waited for days outside clinics offering childhood vaccinations. They had seen the alternative.

So in the 1970s, infectious disease did seem to be on the run. When I was taking medical classes as a research student in the 1970s, my medical colleagues were given Burnet's message not to waste their time on germs. Curing cancer was the future: US president Richard Nixon declared war on it in 1971.

When Peter Piot, now head of the London School of Hygiene and Tropical Medicine, was a student in Belgium, his professors advised him not to specialize in infectious disease. Luckily, he ignored them, helped discover the Ebola virus in the Congo, and later led the global fight against HIV.

Because germs are not gone. As Jeff Goldblum stammers in *Jurassic Park*, life finds a way. Where there are billions of humans to parasitize, some parasite will find us. (Technically, pathogens are parasites, living off the work our bodies do to marshal the energy and tissues that maintain us.)

And the most insidious are the tiny viruses, most little more than a protein shell, maybe with a film of fat, enclosing a clutch of genes, made of either DNA, like our genes, or RNA, the mirror image of DNA we use to translate genes into protein. Viruses carry no energy-capturing or -processing equipment of their own, but use their few proteins to invade and hijack our cells, so they can use them to replicate and spread.

During the twentieth century, we defeated most of the viruses we knew about, mostly with vaccines. However, we didn't realize there were a lot of viruses we didn't know about, which could jump to us from other animal hosts and cause havoc. The buzzword here is "spillover." What Burnet didn't realize about the "wholly unexpected emergence of a new and dangerous infectious disease" was that the fifty years prior to 1972, during which he thought no new diseases had appeared, were no guide to the next fifty.

The first big shock was AIDS, recognized in the United States when gay men started developing rare cancers and pneumonia because their immune systems were suppressed. In 1983, this was traced to the human immunodeficiency virus, HIV, which invades white blood cells of the immune system. By 1984, HIV was found to be widespread in heterosexual people as well, especially in central and eastern Africa.

For a virus that works slowly and is relatively hard to catch—as we all know, it requires the mingling of body fluids—HIV went pandemic shockingly fast. Some forty million people are now living with it worldwide, and it has killed thirty-three million since it was recognized.

HIV shows better than anything else why Burnet's victory whoop was premature. It is a chimpanzee virus that jumped to people around 1920, in southeastern Cameroon, probably when people ate chimp meat or got chimp blood in a cut. Researchers think this sort of viral transfer happens frequently in people who interact closely with animals.

Most such viruses are too ill-adapted to people to settle in and cause an infection, and our immune systems clear them out quickly. A few would have successfully infected us—but when humans were nearly all subsistence farmers, living few and far between in small-ish villages, and rarely traveled, those viruses would have killed a few people, immunized any survivors, run out of victims, and died out in people.

HIV had probably been jumping into the occasional human and not getting much farther ever since its predecessors jumped from monkeys into chimps long ago. But around 1920, the group M strain of the virus hit the big time, when someone carrying it took a boat downstream from Cameroon to the regional boomtown of Leopoldville in the Belgian Congo—now Kinshasa, capital of the Democratic Republic of the Congo.

How do we know all this? In 2014, virologists led by Oliver Pybus at Oxford in Britain and Philippe Lemey at Leuven in Belgium studied some eight hundred HIV viruses from blood samples in old Congo medical records; the oldest was from Leopoldville in 1959. Their genetic sequences differed slightly, showing they had already been circulating among people and acquiring small mutations. Those mutations allowed the team to work out which virus was descended from which and how much time that took, and then construct a family tree. They were all descended from a common ancestor that infected someone around 1920.

Today, Kinshasa is the second largest French-speaking city in the world after Paris, and already in 1920, Leopoldville was no village. It was the capital of the brutal Belgian colonization of central Africa, with fifteen thousand inhabitants. With men pouring in from all over the region to find jobs, the sex trade was brisk. It was also brisk along the rail line to the copper, cobalt, and uranium mines in the southern Katanga region. Tens of thousands of men migrated from around Kinshasa to work in Katanga, and the sex trade followed. The team found the most genetic diversity in HIV samples from Katanga and Kinshasa, meaning the most infections were there.[10]

There was another surge in virus diversity after Congo gained its independence in 1960. Initially, it was largely due to the reuse of needles, a good way to spread HIV. But then wars and upheaval after independence led to a steep increase in poverty and social disruption. Jacques Pépin, of the University of Sherbrooke in Quebec, has calculated that the number of regular customers per female sex worker in Kinshasa shot up from a few long-term regulars to up to a thousand different men a year, leading to a huge surge in infections. Haitians and other foreigners working in Congo left, some with HIV.[11]

So group M HIV went global. It was simply in the right place at the right time—right, at least, for the virus.

AIDS raised some unsettling realizations, and in 1992, the US Institute of Medicine (IOM) issued a widely read report on them. Human numbers were at an all-time high, as was global trade and travel—the word *globalization* had just become common. International disease surveillance was diminishing, just as infectious diseases could travel anywhere more easily than ever. "Profit and liability concerns" had cut incentives for companies to make drugs and vaccines for poor countries, the IOM wrote.

This all added up to a "danger of emerging infectious diseases and the potential for devastating epidemics," they concluded—like the one that had just emerged. Prejudice against the gay men who made up so many of the early cases outside Africa certainly slowed the response to AIDS, unforgivably. But even if that had not been the case, it would still have been true that a previously unknown, horrendous virus suddenly arrived and spread and took the medical world completely by surprise. How many more of those were out there?

Yet the scientific and medical communities, the public, and politicians all seemed complacent, not just about infectious disease in the United States, but globally. "Complacency," the report warned, "can also constitute a major threat to health."

Just because we have suppressed some infectious disease, the IOM wrote, people seem to think we can readily suppress any of them—but old diseases can reemerge, and new ones can emerge. The good news was that we could do something about this. "Anticipation and prevention of infectious diseases are possible, necessary and ultimately cost-effective."[12]

How right they were. The cost of the Covid-19 pandemic runs into trillions of whatever currency you may name, and indeed beyond what can be accounted in money. In 2016, the United States National Academy of Medicine, in a report trenchantly titled "The Neglected Dimension of Global Security: A Framework to Counter Infectious Disease Crises," calculated that dividing the expected cost of future pandemics up into cost per year amounted to $60 billion a year—a figure we might now regard as an underestimate. They figured you could prevent them for $4.5 billion per year.[13] Only now are people other than a few infectious disease experts realizing that it would have been $4.5 billion well spent.

## What Are These Emerging Diseases, and Why Are They Emerging?

Back in 1992, the writers of the emerging diseases report realized that, while all diseases are unique, some features of the AIDS story are typical to most of them. It is human ecology, how we relate to each other and other living things, that drives our illnesses, and especially spillover. And economic globalization, changes in food production, and population growth were all profoundly changing our ecology.

The other important realization was that our infectious diseases mostly start in other animals. Research published in mid-2020 found that the rinderpest virus—a major disease of cattle that was eradicated, after a long campaign, in 2011—gave rise to the measles virus in humans in the sixth century BCE. Flu comes from ducks, smallpox from rodents, malaria from birds, mumps—we think—from pigs.[14,15]

It's no accident those animals are mostly livestock or farm pests. We started living in large numbers and cheek by jowl with animals when we started depending on agriculture, around 10,000 BCE. With crops supplying rich, reliable food, our numbers exploded, and most of us settled near the fields, rather than continuing to wander in small bands of hunter-gatherers.

That exposed us to new dangers. Viruses need hosts. To maintain itself in a human population, a virus needs a constant supply of fresh, nonimmune humans so it can move to a new victim before its current host either dies or develops immune reactions that finish off the virus. That requires a nearby, constantly renewed human population. Measles needs a community of about 250,000 people in order to persist, and in fact cities of that size had just started to appear in western Asia when the measles virus emerged.

Pathogens that persisted in our herds of livestock and other hangers-on had probably been causing occasional disease in people for centuries, but once we started living in our own large herds, a few germs could start permanently circulating in humans. Now we are unprecedentedly numerous and once again providing a promising niche for viruses from a new source: the wild. HIV was a good example, but there are many others, not least Covid-19.

Peter Daszak heads the EcoHealth Alliance, a nonprofit that conducts research aiming to prevent pandemics by promoting wildlife conservation. An Englishman in New York, he became captivated by wildlife diseases in 1995 after discovering a previously unknown pathogen causing diarrhea in a zoo's collection of giant hissing cockroaches. A natural showman, he once carried a pocketful of them into a TED Talk.

Until then, wildlife biologists hadn't been much interested in disease. It wasn't considered important to species survival. They reasoned that as a disease kills a species off, new victims become scarcer, so the disease fails to find new hosts and dies out long before the species does. After the pesticide DDT decimated birds worldwide, chemical pollutants got more attention.

Then, in 1997, a British lab found that a parasite can, in fact, drive a species to extinction on two conditions: it parasitizes more than one species; and one of them tolerates the parasite and keeps it going, even as other host species dwindle and disappear. North American gray squirrels have displaced native red squirrels in much of Europe partly because the grays tolerate the squirrel pox virus. The reds do not.[16] In 2002, the Eurasian West Nile virus exploded across North America, carried by birds and mosquitoes. It got attention by causing occasionally fatal human infections, but it also slaughtered native birds, especially in the crow family, because they had no resistance. But sparrows, originally Eurasian, kept the virus going because they can carry it with no ill effects.[17]

In 1998, Daszak was part of the team that found that just such an effect was allowing a previously unknown family of fungi, chytrids, to cause a massive worldwide die-off of amphibians that had driven some species extinct.[18]

So wildlife biologists started learning about wildlife diseases. Eventually, it became clear that these diseases were also affecting humans. In 2008, Daszak and his colleagues calculated that of 335 novel pathogens that had emerged in people since 1940, 60 percent jumped to us from animals, and 72 percent of those, like Ebola and West Nile, came from wildlife.[19]

The official term for an animal disease that jumps to people is *zoonosis*, from the Greek words for animal and disease. The team also found that the rate at which zoonoses were appearing was rising, as was the percentage of them that came from wildlife as opposed to domestic animals.

Again, the basic problem is our own growing population. When Burnet was writing, there were nearly four billion people in the world. Now there are twice that. More people, and more trade and industry to support them and boost their prosperity, demand more land, minerals, and timber, and more jobs—so people increasingly live and work in previously wild places, mining, lumbering, building, and even catching wild creatures for ever more numerous city dwellers, for uses from pets to medicine. And especially, more people demand more food, so farmers scratch new farms out of forests and turn wild animals into new delicacies. The rest of us crowd into cities along with disease-carrying insects—and other humans.

Daszak and his colleagues mapped where the most new diseases had been observed and found "hotspots" in tropical and subtropical developing countries where economic development was creating concentrations of humans close to many species of wildlife.

This makes sense. The number of species of all kinds increases steadily as you get closer to the equator. There is just that much more, ultimately solar, energy flowing through the system. More species mean more pathogens.

As those species disappear under the onslaught of deforestation or other ecosystem destruction, they at least take their pathogens with them. But in degraded ecosystems, the remaining animals can also carry more pathogens than they might in healthier surroundings, because they are stressed or hungry, and germs take advantage. Also, the animals encounter invading humans more often.

Some biologists suspect a more insidious effect. When pathogens are hosted by several species, some hosts may limit the pathogen's numbers, while others do not. When an environment is degraded, only one kind of host is often left—and those survivors tend to be the "weed" species that live fast, die young, and don't invest much energy in fighting germs. As a result, there might be a greater load of pathogens in the hosts left in a depleted ecosystem than in the original, diverse one.

There has been an increase in Ebola outbreaks since 1994, and researchers suspect they are associated with deforestation, which both displaces and stresses the bats that host Ebola and attracts more humans into bat country. The biggest-ever outbreak of Ebola raged through Guinea, Liberia, and Sierra Leone during 2014, killing at least eleven thousand people. It started in the village of Meliandou in Guinea, where the original dense forest had been largely replaced by cacao, coffee, and other farms.

That left forest bats looking for new homes. Fabian Leendertz of the Robert Koch Institute in Berlin and a team of investigators visited Meliandou after the epidemic, and they found that the village children played in the huge, hollow stump of a rainforest tree near the village, the only remnant of the old forest. A colony of insect-eating bats lived in the stump, of a species that can carry Ebola. The 2014 epidemic started when two-year-old Emile Ouamono managed to contract Ebola, says Leendertz, although he doesn't know if the child was playing with a dead bat as has been reported; his family might have known, but, like Emile, they died.[20]

Most disease surveillance doesn't happen in these high-risk environments, though. It happens where the money and scientists are, in rich temperate countries, even though new diseases, like Covid-19, are far more likely to emerge in these tropical hotspots. There are many hotspots in China, and also India and Indonesia, partly because of large human populations and economic development.

The EcoHealth Alliance says the answer is close surveillance of hotspots for early clusters of disease; research to identify new, potentially zoonotic pathogens in wildlife; and efforts to conserve that wildlife so it will stay healthy and deep inside wilderness areas away from people.

We'll look at how that failed with Covid-19. But for now, let's look at how a pandemic like Covid-19 might happen again if we fail to contain some of the other viruses that are spilling over, that experts think are especially threatening. Some of them are worse than the one we are fighting now.

In fact, disease experts all seem to agree on two things: another pandemic is coming, and no one can predict which pathogen will cause the next one. However, in 2016, the WHO and a panel of scientists decided that some pathogens do bear more watching than others. They made a "blueprint" for R&D (research and development) to equip humanity with vaccines, drugs, and diagnostic tests for the most worrying of these pathogens before they go rogue, and created a list of nine priority viruses for which they figured these should be developed fastest.

The list has already been updated several times, not least to accommodate Covid-19, which was unknown when the scientists made the original selection. To be fair, though, they didn't completely fail to predict this pandemic: the first list did include "coronaviruses." We knew the risk.

Most of the priority pathogens, however, are not previously unknown viruses lurking in wildlife in a hotspot. All but one were chosen because they already cause human disease, and they have been traveling and adapting, which is worrying—especially as we don't have remedies.

Their names, at least, are older and lovelier than Covid-19: Crimean-Congo hemorrhagic fever, Rift Valley fever, Lassa fever, Zika, Nipah, Ebola.

The one exception was dubbed Disease X, an instant hit with newspaper headline writers.[21] It simply means a previously completely unknown pathogen we can't even guess at now, like the ones in wildlife Daszak and his colleagues warn about. It is on the list as a placeholder, for research into ways to respond to unexpected viruses we can't name, such as vaccine "platforms" that can quickly be adapted to create a vaccine for any virus. We'll look at that later. Meanwhile, here are the viral Most Unwanted.

First up, Covid-19, which, whatever else it is, is not Disease X. Coronaviruses were an expected risk and appeared on the list from the start via the two we already knew about, SARS and MERS, and viruses like them. And encouragingly, the list had already done its job of attracting more research and development money to fight them, with a model that proved very handy when Covid hit.

The problem has always been that pharmaceutical companies cannot invest in finding cures for infections that are not a problem—yet—and might never be, no matter how worrying they may look. This is because companies, which do most late-stage drug development, cannot, just by the way they are set up, invest in anything they don't know they can sell for a profit. So the Coalition for Epidemic Preparedness Innovations (CEPI) was set up in 2017 by the Gates Foundation and several governments to fund vaccine R&D for the WHO's priority viruses. Before Covid appeared, CEPI was funding research on five vaccines for MERS, the only severe coronavirus then afflicting humans. That was nearly all the coronavirus vaccine work then happening, but it was still something.

When Covid hit, CEPI's experience allowed it to rapidly organize work on nine vaccine candidates, including the Moderna and AstraZeneca vaccines that were among the first three approved. It also helped arrange government purchasing agreements, alongside GAVI, an agency that arranges public-private funding to help poor countries buy routine vaccines. Both are part of the WHO's COVAX collaboration to ensure poor as well as rich countries get Covid vaccines, one of the pandemic's at least partial success stories.

But CEPI's glass is half full. It is easy to pour money into a real emergency. The worrying part is the limited funding CEPI was getting for MERS and the other nasties on the list before Covid happened. No one was interested in looking at, say, a universal coronavirus vaccine, just one for the only coronavirus already causing severe disease—and thus, a potential market. It will be a measure of whether the world has learned its pandemic lesson if funding for the rest of the priority viruses increases after CEPI's work on Covid is done. It would cost less than Covid has: in 2021 CEPI called for $200 million for that universal coronavirus vaccine, and says $3.5 billion over five years would produce a library of prototype vaccines for twenty-five families of worrying viruses, and massively shorten their development time.

One of those families is the bunyaviruses, with two members on the WHO priority list. Crimean-Congo hemorrhagic fever virus lives in ticks across Asia, Africa, and southeastern Europe. It usually causes just a mild fever in people but can also cause severe disease, killing up to 30 percent of such cases. The European Centre for Disease Prevention and Control (ECDC) says these involve fever, dizziness, sensitivity to light, and "sharp mood swings" in which "the patient may become confused and aggressive."[22]

There is an old Soviet vaccine of unknown efficacy used in Bulgaria, but it isn't widely approved, partly because it is made using mouse brain, which can cause problems. A European research project aims to find a better one—CEPI has two in very early trials.

Meanwhile, the virus is invading new territory as global warming moves the ticks north: it appeared in Western Europe in 2010 and has taken up residence in Spain.[23]

Rift Valley fever is a bunyavirus that mainly infects cattle, but people get it from the cattle via mosquitoes or infected meat. It is found across Africa, but it spread to the Arabian Peninsula in 2000. It, too, is often mild but occasionally causes inflammation and bleeding in the liver, encephalitis—inflammation of the brain—and blindness. Half of those with severe disease die. Livestock can be vaccinated against it, which suggests people can be, too.

Lassa fever, from the arenavirus family, infects five hundred thousand people a year across West Africa, and again, most have mild or no symptoms. However, a few get severely ill, and five thousand a year die. It is carried by the common multimammate rat—yes, that means it has more teats than other rats—so you would think it was unlikely to spread beyond the rat's habitat. But worryingly, it has occasionally shown that it is capable of some human-to-human spread.

Besides, better management of Lassa might help manage other risky pathogens in the same region. In the West African Ebola outbreak in 2014, early cases were initially mistaken for Lassa, which helped the far more contagious Ebola spread. The WHO wants diagnostic tests to fix that. CEPI has six vaccines for Lassa in animal tests.

And there's a further concern with Lassa: it has relatives. In 2008, one that was previously entirely unknown killed a thirty-six-year-old woman in Zambia.[24] At the time, virologists warned how frighteningly little we know about viruses in Africa—where humans have lived longest and where therefore, in theory, more pathogens should be adapted to us than anywhere else. Five of the eight named entries on the WHO's priority list are originally African viruses.

One is Zika, from the flavivirus family that already counts two notorious diseases, dengue and yellow fever. All three are carried by *Aedes* mosquitoes, named with the Greek word for "odious." One notorious *Aedes*, the aggressive tiger mosquito, is migrating outside the tropics with global warming and the global trade in used tires, which harbor puddles where it breeds.

That mosquito migration has already spread emerging disease. Chikungunya, a painful but usually nonlethal *Aedes*-borne virus from East Africa, had a mutation that adapted it to tiger mosquitoes in 2005 and began a series of explosive outbreaks around the Indian Ocean. In 2007 it reached Italy, a rich country that quickly snuffed outbreaks with anti-mosquito campaigns. In 2013, in contrast, it reached the Americas, where many countries could not afford to do that, and it has since caused 2.1 million cases and more than six hundred deaths.[25]

Zika was discovered in monkeys in Uganda in 1947 and then spread in monkeys to Southeast Asia. Until 2006, there had only ever been fourteen known human cases. What it did next was completely unpredicted.

In 2007, Zika caused a big outbreak on the island of Yap in Micronesia, followed in 2013 by French Polynesia and other Pacific islands. The virus's genes showed it came from Southeast Asia, either in an infected human or infected mosquitoes. The insects regularly hitch rides on airplanes, causing cases of "airport malaria" in countries that do not have malaria locally. Zika had been fairly mild the few times it was previously seen in people. But on Yap, some cases developed a potentially paralyzing nerve disorder, Guillain-Barré syndrome, an occasional complication of many different infections, probably caused by wayward immune reactions.

Then in 2015, it appeared in Brazil and rapidly spread across South America and into North America. This time, it was accompanied by severe birth defects in babies born to infected mothers, especially microcephaly, in which the head is abnormally small. Reexamined medical records revealed it had done the same in the Pacific, but no one had made the connection.

Just as they had done with HIV, Oliver Pybus and team at the University of Oxford sequenced the Brazilian viruses and discovered they came from Polynesia. In fact, they were all so alike that, as with Covid-19, Pybus concluded that they probably descended from just one introduction, perhaps only one infected person.[26] That carrier's infection could have resulted from just one mosquito bite. One bite. By late November 2016, Zika had left 3,700 babies in the Americas with birth defects.

Pybus also found that Zika arrived in the Americas in 2013, two years before it was diagnosed—but, like Covid in China before December 2019, at first there were too few cases for limited medical surveillance to notice. It might have happened when Polynesian soccer fans traveled to watch the FIFA Confederations Cup games in Brazil in June 2013.

Zika probably started causing large outbreaks as it left Asia and moved across the Pacific and into the New World because it entered populations of humans who had never encountered it before and had no immunity. Everywhere else it had been, it mostly lived in monkeys but occasionally infected humans. Probably children got it early, had a very mild disease as children can with some viruses, and developed immunity. After one generation of exposure, few adults would be nonimmune. On Yap, there were no monkeys, and no one was immune, so it hit everyone, and adults got a worse disease.

Moreover, sometime before it hit the Pacific, the virus picked up a mutation—just one change of an amino-acid building block in one protein—that helped it invade developing nerve cells and start causing birth defects.[27] There are few better illustrations of one apparently unthreatening microbe suddenly, and unpredictably, causing misery.

As for why Zika suddenly moved east, Pybus suspects it simply started getting more chances: flights from Polynesia to Brazil increased 50 percent between 2012 and 2014. As noted, to spread the virus across the Pacific Ocean, either infected mosquitoes have to make the air journey or an infected human does. For humans to carry it, they have to be bitten by mosquitoes at both ends of the trip: one to give the human the virus, and the second to get it and pass it on to another human.

Both kinds of events are made more likely by the fact that more people than ever before are now flying directly between countries in the Southern Hemisphere, including while it is peak mosquito season in both places. The UN's International Labour Organization reports that, since 2000, the number of people migrating for work within the global south, as opposed to migrating north, has grown from sixty million to eighty-two million a year.

But by 2017, as epidemiologists had predicted, Zika cases in the Americas subsided dramatically, and the WHO ended its declaration of a health emergency. It seems that enough people had been infected to cause herd immunity. That happens when people who haven't had an infection or vaccination, and thus have no immunity, become so rare that it is hard for the virus to reach someone new and susceptible before it dies out in its existing host.

Herd immunity wanes, however, as new people, with no immunity, are born and accumulate, or as the immunity people got from infection wanes. So epidemiologists expect Zika to return. They don't know when—it might be soon, it might be years. It might also migrate elsewhere: two billion people on the planet live with tiger mosquitoes. Experimental Zika vaccines had gotten as far as human safety trials by 2016. But they need to be given to people at risk of catching Zika to test if they work—and there is now too little Zika circulating to test them. Ironically, until it returns, we won't know for sure if we have a working vaccine. This is a perennial problem with emerging diseases.

Meanwhile, managing Zika—and other mosquito-borne illnesses like malaria, chikungunya, and dengue—requires mosquito surveillance and control, and the 2015 epidemic revealed how little capability many health agencies now have to do that. Rich countries that don't have malaria but used to have large antimalaria programs in the tropics abandoned them starting in the government cutbacks of the 1980s, and people and expertise went with them. In 2015, the CDC in the United States found it had only twelve medical entomologists—scientists who can identify and manage mosquitoes that carry disease—to face the Zika invasion. The story goes that it pulled one retiree off a sailboat in the Caribbean to come back to work.

\* \* \*

Then there's the one disease on the WHO priority list that is most likely to keep the scientists awake at night—other than flu, and we'll look at that later. Few people have heard of Nipah virus. Frankly, this one truly scares me.

Scientists who work on emerging diseases meet in Vienna every two years, and in 2016 I went along to talk to them about their research. As we chatted over good Viennese coffee, I conducted a straw poll: Which of these emerging diseases we're hearing about here scares you most? I was expecting the researchers to nominate their own, varied research subjects, but I was surprised. The hands-down winner was Nipah.

If you watched the last scene in the movie *Contagion* carefully, you know about Nipah. The virus is carried by Malaysian flying foxes, the world's biggest bats. Their species name is *vampyrus*, but they are peaceful fruit eaters—with a five-foot, or 1.5-meter, wingspan.

In 1998, forest trees in Malaysia failed to fruit because of a drought caused by the cyclic climate variation El Niño, helped along by global warming plus smoke from human-caused forest fires in Indonesia. That drove big, hungry fruit bats to farms in peninsular Malaysia, including a village called Sungai Nipah, where they ate farmers' fruit—and dropped half-eaten pieces of fruit, as well as urine and feces, into pigpens.

Pigs being pigs, they ate the lot and then developed a severe brain inflammation. People who cared for the sick pigs developed it, too. The virus spread across Malaysia to Singapore, infecting 276 people in the two countries—of whom 106 died. A million pigs were culled in an effort to stop the spread of what was assumed to be a contagious pig disease. Then scientists figured out it was bats.[28]

In 2001, Nipah turned up in Bangladesh and nearby India. It turned out fruit bats were drinking the sweet sap, toddy, that farmers were tapping from palm trees, and contaminating it. The virus now causes a winter outbreak somewhere in the region every year, with death rates up to 75 percent. In 2018, it turned up 1,600 miles away from Bangladesh, across the Indian subcontinent, in the southwestern state of Kerala.

The virus in Malaysia rarely spread person to person, but the Bangladesh virus does, although only to a few successive people before it dies out. The potential to spread more effectively may be there, though, Nipah expert Daniel Lucey of Georgetown University explained to me. Most worryingly, he says, it sometimes causes pneumonia in people and appears to spread in coughed-out droplets. As we all now know, diseases like that can cause trouble.

Kerala is widely considered to have the best public health in India—it also did a good job of flattening its Covid-19 curve in 2020. It isolated and treated people who caught the Nipah virus and contained the outbreak, although seventeen people died, including health care workers—who likely got it from their patients, via person-to-person transmission.

A very similar virus, Hendra, is also carried by fruit bats and has spread to people via horses in Australia. In 2015, an effective, safe vaccine against it was licensed for horses, suggesting human vaccines may be possible. In addition, the University of Queensland has developed a treatment. Antibodies are the proteins produced by your immune system that latch onto a particular pathogen and attract immune cells to destroy it. Various tricks enable immunologists to produce a line of cultured cells that all make the same antibody, called a "monoclonal."

The advantage is that production of a monoclonal antibody can be scaled up by making bigger cell cultures. You usually don't produce your own antibodies to a new pathogen until a week or more after you were infected, and if the initial disease is severe, that can be too late. An injection of monoclonal antibody that attacks the virus can help you fight it earlier. One that attacks Hendra virus—and should also attack Nipah—passed safety tests in people in 2020.

In 2018, the Queensland team sent the antibody to Kerala, but the outbreak was contained before they could use it. Kerala had only one case in 2019, a twenty-three-year-old who recovered, and no cases were spotted in 2020. That doesn't mean the virus is gone, though—in 2020 scientists from the EcoHealth Alliance and colleagues reported that the virus infects fruit bats across India, but the percentage of bats infected rises and falls over periods of a few years, with outbreaks more probable when it rises.[29] Kerala is keeping the antibody in reserve, though we don't know yet how well it will work on Nipah. It will be good to find out: monoclonals are among the most promising treatments for viruses we might be able to brew up in a hurry. Several are being developed for Covid-19 to treat illness and also to prevent infection, potentially a useful backup to vaccination in crowded, high-risk settings like workplaces or nursing homes.[30]

Once again, the perennial problem with developing remedies for emerging diseases is that they are just emerging. Virologists fear Nipah could evolve to cause epidemics, but for now, it strikes only small numbers of people randomly, so tests of therapies are hard to organize. CEPI has four vaccines in animal experiments or human safety trials, but to test whether a vaccine works, it has to be used in an outbreak.

No one knows where one will strike next, but it could turn up in new places. African fruit bats also carry the virus. In 2014, Daszak and his colleagues found that people in Cameroon living in areas undergoing deforestation—and who butchered fruit bats for food—had antibodies to Nipah, showing they had been infected.[31] The Eco-Health team suspects such mild infections might be widespread in India, too, and not diagnosed. This may seem encouraging—if we're missing a lot of mild cases, the true death rate from Nipah might be lower than we think, as we're seeing only the severe cases. That happened in the early days of Covid, too. More worryingly, frequent mild cases could be allowing the virus to adapt to people—and for a virus, adapting means transmitting. We don't want that.

Finally, there is Ebola. The epidemic that raged through 2014 and into 2015 in Liberia, Guinea, and Sierra Leone took the disease world by surprise and was a virtual compendium of the problems the world has in dealing with emerging disease. The virus, also carried by bats, was discovered in the 1970s in central Africa and had previously caused mostly small, containable outbreaks there. No one expected it in West Africa, though it later emerged that the virus had been detected there earlier—but no one had paid much attention. First problem: failure to recognize the risk.

Then it started spreading in Guinea in December 2013 but wasn't recognized as Ebola until March. Second problem: surveillance failure. It continued to spread as Guinea's government initially resisted reporting cases for fear of discouraging foreign investors. Third problem: official secretiveness. The WHO, reluctant to offend a member state and hamstrung by bureaucracy, also dragged its feet.[32] Fourth problem: weak international response mechanisms. When it hit the region's cities, the disease raged out of control, eventually infecting 28,616 people, fifty times more than any previous Ebola outbreak, and according to the most careful observations, killing 70 percent of them. By August 2014, when the WHO declared an emergency, the epidemic curve was exponentially headed for unthinkable heights. Fifth problem: urbanization, especially in poverty-stricken countries with little health care.

Finally, in the nick of time, the world—and the WHO, with a revamped effort led by Bruce Aylward, who later led its Covid-19 mission to China in February 2020—responded and contained the epidemic with the same tools used for Covid-19: isolation, contact tracing, and quarantine. Also, as with Covid-19, changes in ordinary people's behavior were crucial. Friends stopped embracing, and families stopped touching virus-laden corpses at funerals.

As with Covid-19, there were no drugs or vaccines for Ebola: the same market failure that has dogged all emerging diseases prevented their development, even though prototypes were available. After the US anthrax scare in 2001, there was some funding for vaccine research, as Ebola was considered a potential bioweapon. It ended after a few years, however, and there was no money or opportunity to test the vaccine and get it to market.

The West African outbreak changed that—especially after cases appeared in the United States and Europe, and rich governments started paying attention. By early 2015, companies had brewed large batches of the prototypes and pulled off a world's first: testing a vaccine in the middle of a raging epidemic. The experience paid off with Covid.

One of these vaccines was nearly 100 percent effective and would have saved innumerable lives if it could have been deployed faster. But it had taken a year to get it manufactured and approved so it could be deployed in Africa, by which time the epidemic was almost over, and it could only be tested in a few places. That vaccine and another one were, however, deployed in the next Ebola outbreak in 2018 in the conflict-ridden east of the Democratic Republic of the Congo. That epidemic was declared over in November 2020, almost unnoticed amid the Covid-19 pandemic.[33] Despite conflict in the affected region, forty-three thousand people (including nine thousand health care workers) who were exposed to the virus were vaccinated in a massive, successful containment effort. In January 2021, a collaboration of the WHO, UNICEF, the Red Cross, and Doctors Without Borders started an Ebola vaccine stockpile to head off the next outbreak.

The WHO was heavily criticized for its slow response to the 2014 Ebola epidemic. But its role had always been to advise countries on medical treatment, set standards for medical products, and organize long-term efforts like vaccination campaigns. It was supposed to coordinate the response to international outbreaks, too, but it was never designed to be a global emergency response agency. By 2016, in response to the criticism over Ebola, it had undergone a major restructuring to become one. That has stood us in good stead with Covid.

So after dropping the ball on infectious diseases in the 1970s, we have at least been talking about the renewed threat of them for years—as far back as that 1992 report that alerted the world to the growing threat of emerging disease. Yet it took a near-disaster with the world's largest-ever Ebola epidemic in 2014 to reinvent the WHO as the emergency response agency for epidemics we—astonishingly—never had up to that point. The Ebola emergency also launched CEPI and the WHO R&D roadmap, and its pathogens list.

Here's a rare, optimistic thought: think what Covid-19 might inspire us to do.

Yet it should be noted that, as a bat virus still unaccustomed to humans, Ebola spread relatively slowly in West Africa. Researchers have since found, to their horror, that as it spread, the virus adapted to people and may have gotten better at transmitting.[34] That would make it much harder to contain.

Viral evolution is one of the larger unknowns humanity faces as it goes forward into a future where we are all more aware of how vulnerable we are to pandemic disease. We will look at that in more detail later, especially how that might affect some of the viruses on this list, as we turn toward the future.

First, we have enough to handle with the viruses we've got. Let's look at where Covid-19 came from.

# CHAPTER 3

# SARS, MERS — You Can't Say We Weren't Warned

> Experience is that marvelous thing that
> enables you to recognize a mistake when
> you make it again.
>
> —Franklin Jones, twentieth-century US
> journalist, quoted in 2006 by epide-
> miologist Zhong Nanshan in relation
> to SARS[1]

"Have you heard of an epidemic in Guangzhou? An acquaintance of mine from a teacher's chat room lives there and reports that the hospitals there have been closed and people are dying." Stephen Cunnion, an infectious disease expert and former head of preventive medicine for the US Navy, received that email from a friend on February 10, 2003. He couldn't find any further information, so he passed the email on to ProMED.

The same day, ProMED got a notice from the Hong Kong health department warning travelers of a pneumonia outbreak in

Guangdong, the southeastern Chinese province of one hundred million people next to Hong Kong. Guangzhou is its capital and biggest city. ProMED posted both messages.[2] The next day, the WHO asked China about it. The health ministry in Beijing replied that the province did have a pneumonia outbreak that had started the previous November. There had been 305 cases and five deaths.

That was the first the world heard about what was eventually called severe acute respiratory syndrome, or SARS. In the first half of 2003, it traveled to twenty-seven countries, infected 8,096 people, and killed 774, many of them health care workers. Then it was stamped out.

SARS comes up repeatedly in any discussion of Covid-19, as it was in many ways a forerunner. The Covid-19 coronavirus has been officially declared the same species as the SARS virus, named literally SARS-CoV-2. The new virus spreads more readily and is less lethal, but otherwise, it is much the same. And looking back at reporting I did for *New Scientist* about SARS, I'm astonished at how little else seems to have changed.

To understand what unleashed Covid-19 and what needs to happen to prevent the next pandemic, we need to understand SARS. It was, after all, a very clear warning of what is happening now. And then we got two more warnings. And we still did very little.

In February 2003, the WHO already had an inkling something was happening: in late 2002, a Canadian government system that monitors global press reports for mentions of disease picked up reports of pneumonia in China. But the rules in force at the time meant WHO couldn't inquire further about information it had not received officially from the affected government. It couldn't ask China for details until February 11, after Guangzhou made its first public statement about the outbreak.

It had to: locals knew what was happening, and there was panic buying of herbal remedies and vinegar, a traditional disinfectant. The health department said it was caused by a common bacterial infection, mycoplasma, and was under control. ProMED called this "speculative" and posted a press report that many of the people hospitalized were doctors and nurses.[3]

On February 18, the China Center for Disease Control and Prevention (China CDC) said the outbreak was chlamydia, another bacterial infection, and under control. Again, ProMED was politely dubious.[4] Both of the bacteria mentioned so far were good news, in a way, because they could be treated with antibiotics. But both could also be coinfections alongside a primary infection with a virus, for most of which there are no drug treatments. On the 20th, ProMED posted a press report by a foreign-owned news service that quoted a doctor in Guangdong—who did not wish to be identified—saying a virus could not be ruled out.[5]

There was little the WHO could do. The rules at that time were that it could not even tell the world about an outbreak unless the country that had it gave permission. And it could not officially get information about an outbreak from any source but that country's government, leaving it unable to act on information it got anywhere else.

Then our increasingly interconnected world took things up a notch by moving the outbreak across a border. On February 22, Liu

Jianlun, a doctor who had been treating the pneumonia in Guangdong, developed symptoms while staying on the ninth floor of the Metropole Hotel in Hong Kong, where he was attending a wedding. He warned hospital staff to isolate him, knowing what he had, but they had had no official warning of the danger and took insufficient precautions. Some were infected. Liu died ten days later.

Meanwhile, seven other people who stayed on the same floor, plus sixteen more who stayed elsewhere in the Metropole, developed pneumonia—and carried the virus to other hospitals in Hong Kong and to three other countries: Liu was a superspreader. No one ever figured out exactly how they were infected, but SARS traveled in coughed-out droplets like Covid-19. It might have come down to bad luck in sharing an elevator.

Johnny Chen, an American businessman on the ninth floor, developed pneumonia a few days later in Hanoi. His carers also fell ill. Carlo Urbani, a forty-six-year-old Italian infectious disease expert working for the WHO in Hanoi, realized this was a new disease, warned the WHO, and started infection control measures at the hospital. He later developed SARS and died. There is a memorial plaque to him at WHO headquarters in Geneva. The virus cultured from his lungs, known as the Urbani strain, is still used in coronavirus research. It's a macabre but worthy memorial for a scientist who died in the line of duty.

Other people from the Metropole traveled to Canada and Singapore: those countries eventually had some 250 deaths each. Vietnam had 63. Mainland China had 349, Hong Kong, 299.

SARS had a 10 percent death rate, much higher than any estimates for Covid-19 through its first year. It, too, hit the elderly harder but scaled up: half the people over sixty who got it died. Both viruses seem to kill, at least sometimes, by turning on too much of an immune response called inflammation.[6] Normally, inflammation is the way our bodies fight invaders—but in some people, some pathogens can unleash too much of it.

Now that the outbreak was outside mainland China, the WHO could act at the behest of the other affected countries. On March 12, it issued an alert, warning countries and airlines to watch for cases, and citing the outbreaks in Hong Kong and Vietnam—and Guangdong, even though authorities there still officially attributed it to chlamydia. The WHO noted that investigation of the cause in Guangdong was "ongoing." Soon there were cases in Taiwan, Singapore, Thailand, and across the world in Canada. But China still reported only 305 cases—the same number as a month earlier.

At the same time, it asked the WHO for technical assistance. A team arrived in Beijing on March 23. Immediately, China's case numbers jumped to 792, a suggestion that authorities were being more open now with other experts in the room. But the team was kept in Beijing and not allowed to visit the epicenter in Guangdong until April 2.

At that point, things came to a head. The principle of state sovereignty was enshrined in the WHO's founding treaty when it was set up, along with the UN itself, in the wake of World War II, and ruled in disease management as in everything else. The International Health Regulations, a 1969 treaty with antecedents going back to the nineteenth century, prohibited the WHO from doing much without a member state's explicit permission.

But as fears about emerging diseases rose after AIDS and the

1992 Institute of Medicine report, a new idea was taking hold: global health security. The idea was that in a highly interconnected world, diseases could go global fast, so sometimes, for the greater good, an international agency should have the right to interfere in a sovereign state to ensure threatening outbreaks are contained. Implicit in this was the belief—born of repeated experience the world over—that what governments do in their own perceived interests might not be in the interests of the world as a whole, especially when it comes to disease.[7]

In 2003, the director-general of the WHO, Gro Harlem Brundtland, a doctor and the former prime minister of Norway, decided to interfere. Brundtland is still revered by many at the WHO, partly for how she handled SARS. She seemed to have a taste for combat. She once jabbed me in the chest, hard, and then walked away when I asked her at a press reception for her response to some criticism (all she heard was that I had repeated the criticism). I asked one of her aides if she was always like that. "Oh yes," he said.

From the beginning, the SARS virus had demonstrated relatively limited spread in droplets, and all cases could be traced to other cases, meaning containment could work. But on March 30, more than two hundred cases suddenly appeared at the Amoy Gardens apartment complex in Hong Kong. David Heymann, at the time head of infectious diseases at the WHO, told me there were fears it had become airborne, which would make it much harder to control. So on April 2, 2003, Brundtland advised the world to cancel all but essential travel to Hong Kong and Guangdong.

Fears of airborne spread subsided when the Amoy Gardens outbreak was traced to faulty plumbing—like Covid, the virus also spread in feces. But the WHO issued further travel advisories for Beijing and Toronto in late April. None were lifted until each city had largely contained its epidemic, after a week for Toronto but around two months for the others.

It was unheard of for the WHO to issue advice with such direct financial implications for countries without their blessing. The advisories meant considerable lost business: Toronto estimates it lost $265 million. It sent delegations to complain loudly to WHO headquarters in Geneva.

China's health minister reacted more strongly to the travel advisory, publicly urging people the next day to visit Guangdong. Over the next few weeks, authorities limited visits within China by WHO officials and seemed to be understating case numbers. Brundtland was openly critical.

Then on April 9, Jiang Yanyong, a retired surgeon at a Beijing hospital, told Beijing TV stations that claims the epidemic was under control were nonsense, saying there were more than five times the official number of cases just in Beijing. Western media reported it, and Chinese citizens spread the news on cell phone networks.

The next day, Zhong Nanshan, then the head of the Respiratory Disease Research Institute in Guangdong, told the press, "The origin of this disease is still not clear, so how can you say it has been controlled?"[8] He suspected a virus and later wrote that, in fact, a Chinese lab had identified the coronavirus as early as February 26 but kept it quiet, as official sources blamed bacteria.

Viral sequencing was slower and more expensive in 2003 than it was when Wuhan's hospitals sent their mystery virus for sequencing in 2019. But on April 14, a Canadian lab reported that it had sequenced a coronavirus from a SARS patient. Further claims that bacteria were responsible were no longer credible, and the WHO confirmed the virus as the cause of the pneumonia.

Yanzhong Huang of Seton Hall University and the US Council on Foreign Relations, who has researched the period, writes that on April 17, the top committee of the Chinese Communist Party called for a change in policy, and China's president, Hu Jintao, instructed officials to stop withholding information about the epidemic. Two days after that, officials admitted that Beijing had 346 SARS cases. The reported number had been 37.[9]

Outside China, the world had swung into action. Heymann set up daily conference calls between doctors, epidemiologists, and virologists around the world, to compare notes on the best treatment, speed up the development of tests to diagnose the virus, and work out how it spread. The worldwide conference call became a durable model, Heymann told me in March 2020: the same groups were convened for Covid-19, this time online.

Then, as now, what was crucial was old-fashioned epidemiology. Cases were isolated. Their contacts were traced and quarantined. Workers in affected hospitals self-isolated if they developed a fever. Hong Kong barred contacts of cases from leaving the territory, and police went after quarantine breakers. Singapore monitored quarantine with then-novel internet cameras. Toronto eased up on quarantine too soon and almost lost control, but got it back.

However, at the time, senior health experts told me that while this containment might slow the epidemic, the virus would inevitably get into cities that simply did not have the resources or the social order required to contain it. SARS was here to stay. On April 26, I wrote in *New Scientist* that rich countries, in pure self-interest, must ensure that any eventual vaccine went to both rich and poor—because "sooner or later, SARS is coming to a person near you."

But my sources were wrong, even though their fears were entirely reasonable. Sometimes, luck takes a hand: the virus never reached a chaotic megacity like Kinshasa or Calcutta. On July 5, there had been no new cases for three weeks, and the WHO declared SARS "contained."

What had also happened was that after abandoning its efforts to conceal SARS, China launched mass mobilization to contain the virus, confining Beijing students to dorms and spending more than $1 billion refurbishing hospitals and finding and isolating cases. Sound familiar? The abrupt shift from downplaying the outbreak to a no-holds-barred response was eerily similar to what later happened with Covid-19. Then, as now, it worked. And it would have been a lot easier—and lives around the world might have been saved—if it had happened earlier.

Since then, more of the underlying story has come out. Huang wrote in 2004 that Guangdong health officials initially recognized the new disease as a virus and alerted authorities—but the law made any infectious disease outbreak a state secret until the health ministry announced it, so they couldn't tell anyone else. Then there were bureaucratic delays at the ministry, some because of the Lunar New Year holiday, until February 11.[10]

Then the news blackout resumed during the National People's

Congress in March, just as it did during the Party Congress in Wuhan with Covid-19. Lower-level officials soft-pedaled reports to senior officials, Huang found, for fear of looking bad. Again, there are parallels, he says, with Covid-19.

Having faced a harrowing near-miss on SARS, you'd think we'd be doing better by now. And in fact, in the aftermath, China installed the automated alert system we talked about in Chapter 1, meant to allow doctors to alert central authorities to certain illnesses, notably undiagnosed pneumonia, bypassing bureaucrats to make sure the blockages that delayed reporting SARS didn't happen again. Yet it was sidelined when Covid-19 emerged, by the same bureaucratic culture of suppressing bad news.

There were a few quickly contained outbreaks of the virus the year after SARS was beaten, including escapes from virology labs and some outbreaks described as having been from a "wild source." Both types were troubling.

Labs remain a worrying source of dangerous viruses, although scientists and regulators tightened up precautions after the SARS incidents. In April 2020, some alleged that the Covid-19 virus might have escaped from the Wuhan Institute of Virology, where China's first highest-level containment lab, the kind used to study the most dangerous pathogens, opened in 2015. No evidence for this was offered. But there are concerns with all these labs that despite the showers and filters and hazmat suits, a scientist might be infected with a virus that sometimes causes few symptoms, and then carry it outside.

And the possibility remained that the SARS virus—or something very like it—still lurked in the wildlife it had presumably come from in the first place. That, in fact, was discovered by the virologists at the Wuhan Institute, initially in 2005 and definitively in 2017, and they warned the world about it. We'll discuss that later.

But although some SARS virus remained in a few lab freezers and wildlife, it was clear that SARS was gone from people: a triumph, if a close shave. What was crucial in controlling SARS, says Heymann, was that, unlike Covid-19, the virus could not be spread by oral or nasal droplets until late in the infection, well after symptoms started, because it did not build up in the nose and throat until then. Thus, if you isolated every exposed person with a fever, you had contained it. With Covid-19, people with symptoms have already been spreading virus around for a day or two. Viruses that spread before they cause symptoms are very hard to contain: look at HIV.

Because SARS didn't spread as readily as Covid-19, no severe social distancing was needed to slow its spread enough to cut the number of contacts that had to be quarantined and make containment possible. There also seem to have been no asymptomatic cases. So SARS never spread far in the community.

And as we saw, it never got into any big cities in poor countries that could not have contained it. That did not happen this time: the number of flights from China to such places has increased tenfold or more since then. That is partly due to increased travel globally and the vastly increased prosperity of many Chinese. Contacts have also increased due to the Belt and Road Initiative, China's massive worldwide program for investment and infrastructure.

Simply put, in 2003 we dodged a bullet: ProMED, WHO leadership, global collaboration among experts, shoe-leather epidemiology, and, eventually, massive Chinese action—led finally by its doctors and scientists—eradicated SARS. They were helped by the fact that the virus was simply clumsier at spreading among humans than Covid-19.

What stands out now, in retrospect, is the speed and efficiency with which the world acted. The virus didn't get opportunities to establish itself in countries outside China that delayed taking action to contain it, as happened with Covid-19. As the virus arrived, there was no disputing the need for containment or talk of relying on herd immunity. Because of this swift action, experts still debate whether, despite its global transmission, SARS ever circulated widely enough to be called a real pandemic.

Maybe the virus's high death rate scared everyone into line. Maybe its inability to spread before symptoms started, and absence of many mild cases, just made it easier and less disruptive to follow the epidemiologists' advice. Some sociologists think there was more public trust in experts in 2003.[11]

But did we learn our lesson with SARS and apply that to its sibling, Covid-19? In its World Health Report in late 2003, the WHO listed the seven top lessons from the epidemic.[12]

Lesson seven was that risk communication about a new disease is hard, but vital. When Covid hit, the WHO and countries that successfully contained the virus made it a priority. Others, notably the United States, suffered from what the WHO dubbed an infodemic of misunderstanding and outright lies, from official and unofficial sources. Lesson partly learned.

Lesson six: even without drugs and vaccines, old-fashioned "nonpharmaceutical" measures like quarantine and contact tracing, plus global consultations among all the countries affected, can contain a disease. Yes, but it helps when the virus cooperates like SARS did. It helps even more if governments don't imagine they can take shortcuts around the epidemiology to avoid the economic or political impacts of those old-fashioned measures. Sweden, the UK, the United States, initially China . . . lesson partly not learned.

Lesson five was that health systems should protect health care workers, who made up between one-third and two-thirds of SARS cases in heavily affected countries. Most nurses were (and are) women, and among health care staff, women were 2.7 times more likely than men to get SARS, whereas outside of hospitals, it infected both equally. Yet now, even in some rich countries, doctors and nurses have died in unconscionable numbers because they had to face Covid-19 with insufficient masks, gloves, and gowns. Lesson not learned.

Lesson four: "The world's scientists, clinicians and public health experts are willing to set aside academic competition and work together for the public health good when the situation so requires." To say that has happened again with Covid-19 would be a massive understatement: the outpouring of scientific and medical collaboration has been astonishing, as has the sheer quantity of research, posted almost before the ink is dry on the data, on preprint servers like bioRxiv or medRxiv. This also means it's posted before the usual expert reviewers have formally decided it's okay, which has allowed some less-than-stellar science to be posted—but in most cases, other scientists piled on with reviews on social media (of varying politeness) that prevented it from having undue influence.

This has been one of the silver linings of the pandemic. "I'm

constantly amazed at how easily all the technical people communicate," said Bruce Aylward of the WHO, after returning from
investigating Covid-19 in China in February 2020. For me covering
the story, it's been amazing watching a global scientific community
deal 24/7 for months with a truly global crisis. And with Covid as
with SARS, it was doctors and scientists, in China and elsewhere,
who told the world how serious things were when politicians did
not. I'm not sure we had to learn this lesson; we already knew.

Lesson three was that travel restrictions can help, although the
WHO conceded that temperature checks in airports caught only
two cases of SARS. This one is tricky. Both modeling and observation have suggested closing borders doesn't always achieve
much. The refusal of airlines to fly to countries affected by Ebola
in 2014—mostly out of insurance concerns—severely impeded the
international response.

And sometimes border closures can hurt more than they help.
The bacterial disease plague, linked to the medieval Black Death,
which is slow-spreading and treatable with antibiotics, struck Surat,
India, in 1994. It ultimately killed only fifty-six people, yet India lost
$3 billion in trade and tourism when international panic over the
outbreak closed borders to travel and trade with India. Hence when
the International Health Regulations, the treaty that governs international epidemics, were revised in 2005, they discouraged countries from closing borders as a knee-jerk response to epidemics,
partly so countries wouldn't conceal outbreaks for fear of losing
money. Accordingly, with Covid-19, the WHO advised against it.

Few listened. Countries the world over closed their borders
to slow the spread of Covid-19. This didn't always make sense

epidemiologically and sometimes, as in the United States, gave a false feeling of security that hampered response to the virus already spreading within the country. But in many cases travel restrictions helped, notably with the initial massive response in China, with islands like New Zealand, and within the EU—even within single countries in the EU. Let's say lesson learned, even if the WHO itself has been reluctant to learn it.

Lesson two was that global alerts work. After the WHO issued its alert on SARS in March 2003, affected countries redoubled efforts and got the epidemic under control, and others kept imported cases from spreading. The International Health Regulations treaty was massively revised in 2005 because of SARS and enshrines this by having WHO declare a Public Health Emergency of International Concern when an extraordinary threat looms. With Covid-19, they made that declaration on January 30. Lesson learned.

Lesson one is worth quoting in full:

The first and most compelling lesson concerns the need to report, promptly and openly, cases of any disease with the potential for international spread. Attempts to conceal cases of an infectious disease, for fear of social and economic consequences, must be recognized as a short-term stop-gap measure that carries a very high price: the potential for high levels of human suffering and death, loss of credibility in the eyes of the international community, escalating negative domestic economic impact, damage to the health and economies of neighbouring countries, and a very real risk that outbreaks within the country's own territory will spiral out of control.... Strengthening of systems for outbreak alert and response [is] the only rational way to defend public health security against not only SARS but also against all future infectious disease threats.

It's painful to read that now. And the WHO meant it to apply to all countries. Even though the burden of sounding the alarm fell to China in 2003 and again in 2020, the whole world needs this lesson. China certainly was more open about many things from the start on Covid-19 than it was about SARS in 2003—except for the crucial detail of it being contagious. Lesson very much not learned.

So that's two learned, two not learned, two partly learned, and one not needed. Clearly, the lessons unlearned from SARS have been deadly.

Case in point: I wondered in *New Scientist* in April 2003 whether, if Hong Kong hospital staff had known more about this new pneumonia in February, they might have used better infection control and not let it spread further. Or if China had acted sooner, whether it could have limited SARS to Guangdong. We're wondering those kinds of things now about Covid-19.

If nothing else, though, you would think we would have made a point of developing some coronavirus drugs and vaccines after SARS, just in case it or something like it ever came back, which, of course, it now has. But China's bureaucracy is not the only system failure in this saga of global disease mismanagement. Western capitalism has its glitches as well.

The silver cloud of eradicating SARS had a dark lining. Vaccine labs and antiviral drug developers set to work as soon as the SARS virus was identified, and their findings are being dusted off now to fight Covid-19. But, those experts say, funding to continue the research dried up after 2005, precisely because SARS had been eradicated—so our knowledge and tools are nowhere near what they could have been if that work had continued.

With no virus circulating, it is, if nothing else, hard to test whether a drug or vaccine works, because normally you do that by treating infected people or by vaccinating people and watching to see if they get infected. You can look for alternative measures of success or failure, such as durable immune reactions to a vaccine in humans or convincing results curing (with drugs) or protecting (with vaccines) animals exposed to the virus experimentally, in a high-containment lab.

But no one bothered doing this, says David Heymann, because with no SARS virus circulating, there was no market for any SARS drugs or vaccines they developed. Only big pharmaceutical firms have the know-how and money to get a drug or vaccine through the big, complex safety and effectiveness trials they rightly need before governments approve their use. Without a market, as we noted in the last chapter in connection with the launch of CEPI, the companies cannot invest in these expensive trials because they don't know that they will ever make their R&D investment back by selling the eventual products.

At one time, as we saw with smallpox vaccines, some pharmaceutical companies were state-owned and could undertake work for the public good. But since the 1980s, those have disappeared, and pharmaceutical development is all done by private companies that are required to turn a profit. It's not because they're mean, it's because we decided as a society to do it that way, influenced by ideas that as much as possible should be done by the market rather than government. With SARS gone, it was too big a financial risk for a private company to invest in SARS drugs and vaccines. There was no guarantee they would ever be needed.

The same market failure stifles R&D for other vital medicines that for various reasons cannot be sold in large enough quantities or at high enough prices to make the R&D investment back, most worryingly new antibiotics. Mechanisms for getting around this problem, by rewarding drug developers in ways not tied to selling the product, have been discussed, but little has been attempted on a commercial scale.

This all makes developing products for any emerging disease difficult or impossible for the private companies that do much of our drug and vaccine development. However, public good is making a comeback, and was even before Covid. Over the past decade, public-private partnerships have emerged to develop drugs and vaccines for diseases mostly found in poor countries, funded by the Bill and Melinda Gates Foundation and others. CEPI, launched just in time for the Covid crisis, organizes such funding for vaccine R&D for emerging diseases, and now for Covid-19.

But there was another reason no real effort was made after SARS to develop remedies for coronaviruses: some virologists decided SARS was never coming back. They made two mistakes. Rolf Hilgenfeld, of the University of Lübeck in Germany, was working on anti-coronavirus drugs that block an enzyme the viruses need to infect cells, when funding for that research dried up in 2006 —development of such "protease inhibitors" has now resumed. He says one mistake was based on a major genetic difference between SARS and other coronaviruses: in SARS, a swath of twenty-nine nucleotides found in other coronaviruses was simply missing from one gene.

Such "deletions" are certainly not unknown in viruses like these, which keep their genes as RNA rather than DNA: RNA viruses tend to be less genetically stable. And the gene's function was then unknown. Yet, some argued this major change was what enabled SARS to suddenly spread in humans. And, they reasoned, exactly the same dramatic mutation was unlikely to happen again. Ergo, SARS was not coming back.

Other virologists disagreed. "I certainly never said that," says Ab Osterhaus, a leading virologist, whose lab did experiments in 2003 that clinched the proof that the SARS virus caused the disease. Those who did say it were at least right about the deletion not coming back. The Covid-19 virus did not have it, at least when it emerged. But Covid-19 spreads even better in humans than SARS did, so clearly the mutation didn't play the role some imagined.

But there was also a second mistake: thinking SARS was gone in wildlife, too. In 2005, the repeated failure to find the virus in civets led some researchers to conclude the virus was gone in nature and had disappeared as a threat.[13]

People who fell ill with SARS were initially linked to a wildlife market, as were some of the first people with Covid-19. The SARS virus was found in Guangdong markets in cages that had been occupied by masked palm civets, a member of a family of mammals related to cats, which in China are bred on farms and sold as game meat. TRAFFIC, an environmental group headquartered in Cambridge, England, that monitors trade in endangered wildlife, says some ten thousand civets were destroyed in markets in China in 2003 in an effort to stamp out the virus.

Tragically for those civets, virologists now think they had little to do with it. The virus came from bats. A few civets and another mammal, the raccoon dog, in Guangdong markets were found to have SARS or to have been infected with it in the past, and attention focused on civets because far more of them were sold. But, said virologists reviewing the research in 2007, SARS was never found in civets anywhere else, wild or farmed, and the evidence suggests the animals were infected at the market by something else, the same as humans were.[14] Nonetheless, the story that civets were the "intermediate species" that spread the virus to humans became widespread. Similar tales are now being told about Covid-19 and pangolins.

In 2005, though, Chinese scientists were already warning that the virus could well lurk in other species. Virologists at the Wuhan Institute of Virology reported coronaviruses that were very similar to SARS in bats, which were also sold at markets.[15] "If no action is taken to control wildlife markets," Zhong Nanshan warned in 2006, the SARS virus might once again "develop into an epidemic strain."[16] And if it did, we would need drugs and vaccines. Nevertheless, companies and government research funding agencies went with the more welcome assessment that SARS was gone.

And yet, between the disappearance of SARS from humans and the arrival of Covid-19, we got another warning about these coronaviruses: MERS.

In June 2012, Ali Zaki, an Egyptian virologist working at a hospital in Jeddah, Saudi Arabia, couldn't identify what had killed a sixty-year-old man with pneumonia. The only positive test was a generic one for coronaviruses. But SARS was gone, and the other known coronaviruses in humans cause common colds. If any virologist could identify an unknown virus quickly, Zaki thought, it's Ron Fouchier in Rotterdam. He sent him samples.

Normally, scientists can't afford to take time identifying odd viruses that crop up—"stamp collecting" they dismissively call it. You rarely get a publication out of it, and researchers' jobs depend on getting research grants, which are given for research that does result in publications.

But one good response to SARS was a program in the EU that funded researchers to do the odd stamp collecting for mystery diseases—just in case they discovered something important. Fouchier had funding from the program, and he discovered a previously unknown coronavirus in Zaki's sample. Worryingly, it was, like SARS, closely related to what by that time virologists knew were bat coronaviruses.

Zaki posted it on ProMED.[17] A British hospital immediately discovered the same virus in a man with undiagnosed pneumonia who had just been in Saudi Arabia.

Within days, Zaki told me later, the Saudi health ministry had sent an "aggressive" and "threatening" team to investigate his lab. He took emergency leave in Cairo. He was sacked—and informed, he told me, that it was not safe to return to Jeddah.[18]

The Saudi deputy health minister, Ziad Memish, told me it was intolerable that Saudi authorities didn't know about the virus until they saw it on ProMED, three months after the patient died—and just as preparations were at their peak for the biggest annual human gathering on earth, the Hajj in Mecca. That's a real concern: Memish helped run the exactingly thorough Saudi health controls aimed at stopping anything worse than the ubiquitous "Hajj cough" from marring the pilgrimage.

Zaki and Fouchier, though, told me that the upcoming Hajj made it a good thing the virus had been identified so fast, as that allowed them to find more cases and establish that it did not spread readily. Zaki was convinced that this would not have happened so quickly if he had told only the Saudi authorities.

The virus was named Middle East respiratory syndrome, MERS, as cases were soon found all over the region, including a hospital outbreak in Lebanon. It was in bats, but people got it from camels.[19] Increasingly intensive camel raising in the Arabian Peninsula is thought to have made the infection endemic in the area's dromedaries.

As of March 2021, there had been 2,586 cases worldwide, four-fifths of them in Saudi Arabia, and 939 deaths—this virus has a high death rate. In 2015, a man who had been in the Arabian Peninsula brought MERS to South Korea, starting an outbreak mostly within hospitals: 184 cases, 38 deaths. MERS has appeared in twenty-seven countries, usually as only one or a few cases after someone has returned from travel to the Middle East.

That isn't much spread in eight years. The main reason is that the virus has never quite made itself comfortable in people. MERS can spread from one person to another, but chains of infection die out after a few cases: epidemiologists call it "stuttering" transmission. Cases keep happening, though, because people keep catching it from camels and starting short chains of transmission. In viruses that jump to us from animals, contending with our entirely new immune system can be too difficult, and the viruses that manage to get out and into another person might be too few to get much farther.

Moreover, unlike SARS and Covid-19, which latch onto cell-surface proteins in the nose and throat, MERS binds to proteins mostly in the deep lungs. That is one reason it is more lethal than the other two viruses—infections there can kill you. But it also means the virus literally has trouble getting out and into the next victim. Your deep lungs don't cough and sneeze.

The worry with such a virus is that it is under enormous pressure to adapt to its new host, and if MERS does, we could get a virus that spreads more readily but is also deadly. To prevent that, you want to keep its opportunities to infect people, and adapt, to a minimum. This especially applies to protecting patients and staff in a hospital facing a MERS outbreak: even more than SARS, this is a disease of health care, partly because some medical procedures, such as inserting a ventilator into someone severely ill with pneumonia, can expel it from the deep lungs and into someone else. That has happened with Covid-19 as well.

Hospital infection control staff made huge strides in restricting the spread of MERS, first in Middle Eastern hospitals, then after it invaded South Korean hospitals in 2015. In 2019, epidemiologists calculated that increased efforts to diagnose and contain the virus early had averted up to five hundred cases since 2016.[20]

So unless it evolves, MERS isn't threatening most of us any time soon. But it is worth knowing about for three reasons. One, it shows that China isn't the only country that doesn't like nasty new diseases being reported on its territory, or involving foreigners in the response. In years of covering infectious disease, I have encountered all too many examples. BSE (bovine spongiform encephalopathy), or mad cow disease, started in Britain, which initially downplayed it. Then, despite science showing it had to be in cattle in Continental Europe, countries there denied it for years—even though we knew by 1996 that it caused a devastating disease in humans. When I reported that science in 1997, there was uproar in Belgium, with questions asked in Parliament and leading veterinary scientists forced to lie to support the official denial.[21]

Two, as the only threatening human coronavirus left in circulation after SARS was stamped out, MERS was actually the subject of some coronavirus vaccine work when Covid-19 hit. A good MERS vaccine for domestic camels should stop it getting into humans. The work on MERS led by CEPI helped in the response to Covid, by showing that the coronavirus's spike protein makes a good vaccine—that proved true for Covid as well. Laudably, CEPI's MERS work continued in addition to its Covid work in 2020, despite the pandemic emergency.

Three, if SARS wasn't enough, the appearance of MERS definitely demonstrated that we should have been working more urgently to prepare for coronavirus outbreaks. How many warnings did we need?

In fact, we got a third one. In 2016, piglets started dying on farms sixty miles from Foshan, the town in Guangdong that was home to the first SARS cases in people. It was called SADS: swine acute diarrhea syndrome. Virologists isolated a coronavirus and found it was 98.5 percent identical to one found in the droppings of horseshoe bats in a nearby cave, the same species linked to SARS and Covid-19.

The pigs had probably eaten some droppings. Nearly twenty-five thousand piglets died, and the disease broke out again in 2019.

There weren't any infections among the farmers. But in 2019, scientists at Zhejiang University in Hangzhou found that SADS can infect cultured cells from humans. So here was another bat coronavirus, it was killing mammals, and it might well infect people. Yet we hadn't done much to guard against these things until we were locked in a global struggle against Covid-19.

Except for a woman named Shi Zhengli at the Wuhan Institute of Virology, and the EcoHealth Alliance. They've been chasing coronaviruses to where they live—in bats. And that could hold the key to finally getting a grip on these viruses.

CHAPTER 4

# Don't Blame the Bats

> We have met the enemy, and he is us.
> —*Pogo*, a comic strip by Walt Kelly

The Covid-19 virus comes from bats. So did SARS. So do MERS, Ebola, Marburg, Nipah, Hendra, and Lassa viruses. So does hepatitis C, which an estimated seventy-one million people are living with worldwide.[1] After biting the head off what he had assumed was a rubber bat a fan threw onstage during a 1982 concert in Iowa, heavy metal singer Ozzy Osbourne needed the long, painful series of injections then required to prevent rabies—another bat virus. (Today the treatment is a little easier.)

And those are just a few of the viruses living in bats that we know cause human disease. In April 2020, researchers reported six kinds of coronavirus previously unknown to science in bats in Myanmar.[2] That adds to the 400-odd found already in China. In November 2020, researchers in Cambodia and Japan found viruses closely related to SARS-2 in bat samples in lab freezers.[3] In 2017, a survey of all known gene sequences of coronaviruses found a hundred "clusters," essentially family groups, of the viruses. Ninety-one of those live in bats, making bats the world headquarters of coronavirus evolution.[4] And they carry other kinds of viruses, too.

If we want to understand this pandemic and what we need to do to stop the next one, the connection between bats and viruses needs to be explored, for three reasons. One, we're going to have to work out why bats have all these viruses. Two, we must learn which of their viruses might jump to us and take measures to prevent and prepare. And three, and most importantly, we must learn how to act, globally, on this kind of information, or there will be no point getting it. We had it for Covid-19, and we didn't use it.

That's right. A lab in China, in work confirmed by virologists in the United States, found a virus very similar to the one that causes Covid-19 in bats in 2013—a full seven years before this pandemic swept the world. Both Chinese and American scientists clearly warned that this kind of virus could well cause a pandemic. They even found bat viruses already capable of infecting human lungs, no further genetic changes required. And yet no serious action of any kind was taken. It was no one's job to do that. This is one of the major things we need to change.

We knew bats in the Americas carried rabies in the 1950s, but no one knew they harbored this plethora of viruses until 1994. That year, flying foxes, a kind of fruit bat, were found to carry a mystery virus that had killed horses—and two of their human caretakers— in Hendra, a suburb of Brisbane, Australia. After that, the more virologists looked, the more they found.

Wildlife scientists, fearing bats would be persecuted because of this, have accused the virologists of disproportionately targeting bats for viral discovery.[5] But a 2017 review of the research showed that, even after accounting for different amounts of research effort, bats were still significantly more likely to harbor diseases affecting humans than any other group of mammals.[6]

Soon after the SARS epidemic took hold in 2003, Chinese scientists started the long hunt for the source of the virus. As we saw in Chapter 3, the SARS virus was initially found in masked palm civets at a wet market—but only in Guangdong, never in any other wild or farmed civets. In fact, civets infected with SARS got sick, making them unlikely to be the home of the virus in the wild, where sick animals don't last long.

In 2004, Shi Zhengli of the Wuhan Institute of Virology and her colleagues started looking for the SARS virus in nature. The team wondered if both the Guangdong civets and people got it straight from whatever animal was the real "reservoir" for the virus, the term for a species that can carry a virus and pass it on but does not itself get sick from it.

Shi's team knew bats could carry viruses without getting sick— and "bats and bat products," they wrote, were increasingly turning up as food or traditional medicine in markets in southern China. So they went to bat caves across China and took blood, urine, feces, and throat swabs from dozens of bats of various species. Shi was nicknamed "bat woman" by her colleagues.

Sure enough, there were viruses 94 percent identical to SARS in insect-eating horseshoe bats in Hubei province, where Wuhan is. The same bats live in many other Chinese provinces and across

Eurasia. The bat viruses were all similar, but had between them more slight genetic variations than the SARS viruses from humans or civets, although those fit within a family tree of all these viruses. That is what you'd expect if bats are where these viruses live naturally and just one or a few of them moved into civets and us.

None of the viruses they initially found was exactly the same as the SARS virus. For one thing, none had the same binding area on the big spike proteins on the outside of the virus, which in SARS latches onto the ACE2 protein on human (and civet and bat) cells—the same receptor Covid-19 uses.[7] But they kept looking.

They also partnered with the PREDICT program of the US Agency for International Development (USAID). PREDICT was launched after the H5N1 bird flu scare in 2004, which we'll look at later. It sets up local labs and surveillance in countries with "hotspots" of zoonoses, and the EcoHealth Alliance is a major participant. Kevin Olival, now EcoHealth's vice president for research, mostly works in Indonesia and Thailand, countries that can use more help building virology infrastructure than China, one of the world's leading producers of world-class biomedical research.

But a PREDICT team worked until recently alongside Chinese scientists at a forest site in China's southern Yunnan province. It is only forty miles from Kunming, a city of six million, and it is in a zoonosis hotspot, with a well-populated bat cave. Even though the same bat species live elsewhere in China, the team has focused on Yunnan as it allows them to readily sample a lot of bats in one population. It is also a poor region where people are more likely to encounter bats and maybe become infected, so research there seems especially useful. Olival told me about it.

The procedure, he said, is to trap bats when they flutter out of the cave to hunt just after dark. The trap looks like a giant harp, with two sets of vertical cords strung in an open frame. The echolocating bats detect the first cords and do a mid-air turn to fly between them—only to become lodged against the second set. No longer able to stay aloft, they slide down into a big, soft bag at the bottom. "They just snuggle up in there," says Olival.

The scientists and technicians already have lights, bottles, labels, and sampling swabs laid out on a folding table nearby. They take a throat swab, an anal swab, and a blood sample from each bat—then the bats fly off to resume hunting. As a conservation as well as disease-hunting organization, says Olival, "we don't want to harm the bats."

The samples from Yunnan are analyzed for coronaviruses, but PREDICT has operations like this looking at different kinds of high-risk wildlife, and different pathogens, in Bangladesh, Brazil, Colombia, Indonesia, Malaysia, and Mexico as well. The findings are analyzed and mapped to predict potential disease outbreaks, and importantly, says Olival, the information is fed back to the communities at risk, so they can protect themselves.

The collaboration with Wuhan quickly paid off. In 2013, Shi's lab found two viruses in the Yunnan bats that were 95 percent identical to SARS and had an exterior spike protein with a sequence they knew would bind to the same ACE2 protein on human cells that SARS used to invade us. The two sequenced viruses were dead—the process of extracting them from the bat samples kills them, which also makes them less risky and easier to deal with.

But they also brought some samples back alive. From one, the team managed to isolate a live coronavirus that could infect both bat and human cells. It was also instantly recognized by antibodies, immune proteins that are highly specific for particular pathogens, taken from SARS patients in 2003. "Bat coronaviruses remain a substantial global threat to public health," Shi's team concluded.[8]

In 2017, the team reported more SARS-like viruses from the bats and discovered that, like some other viruses, they swap gene segments around. They found segments with all the exact gene sequences required to build the original SARS virus in bats from the one cave near Kunming and evidence that the viruses were actively recombining gene segments. After fourteen years, the long search was over: they knew for sure where SARS came from.

But besides SARS, they found a wide variety of slightly different coronaviruses similar to SARS that are able to latch onto the human ACE2 protein to invade cells. "The risk of spillover into people and emergence of a disease similar to SARS is possible," they warned.[9] That, of course, is what has now happened: Covid-19 is a SARS-like bat coronavirus that latches on to ACE2.

Meanwhile, virologist Ralph Baric and his team at the University of North Carolina reconstructed one of the viruses Wuhan had discovered using the gene sequences and found it infected human airway cells in culture just as well as the SARS virus—the Urbani strain. It made mice with human ACE2 proteins, effectively model humans, sick. Yet it was different enough from SARS that an experimental SARS vaccine didn't protect the mice—showing that even if we beat one type of coronavirus, even very similar ones could be entirely new challenges. The title of the 2015 publication of the work warned that SARS-like bat coronaviruses "show potential for

human emergence," and the report spoke of "a need for both surveillance and improved therapeutics against circulating SARS-like
viruses."[10]

In 2016, the title of another paper from the lab called one of
the viruses "poised for human emergence." "The virus has significant pathogenic potential," Baric and the team concluded—and if
it emerged, we had no vaccines.[11]

So we knew there were viruses like SARS out there that could
infect humans and cause illness without having to spend time
adapting in another "intermediate" species first—they could infect
us straight from the bat. We knew this seven years ago. And since
then, more research has only confirmed that.

It was even reported in the press: at that meeting in 2016 in
Vienna on emerging diseases where people were worried about
Nipah, Kevin Olival told me, and I reported, that PREDICT had
helped find "a Chinese virus closely related to SARS but different
enough that prototype SARS vaccines won't work against it."[12]

It got even more alarming. In 2018, Shi's team reported that the
viruses were already trying us out. They found antibodies to the
bat coronaviruses in people living near the caves in Yunnan, showing they had been infected—and they had not been exposed to
SARS in 2003, or traveled. There were also antibodies to SARS-like
viruses in market traders in Guangdong in 2001, well before SARS
exploded. But those weren't found until 2004, after SARS was gone,
in a retrospective analysis of stored blood samples. With SARS we
were caught blind, but with Covid-19, we knew these viruses were
checking us out before this pandemic even happened.[13]

And we knew that due to China's size, dense population, and

geographic variability, it has a large diversity of bat species, potentially making the risk especially great in China. "It is highly likely that future SARS- or MERS-like coronavirus outbreaks will originate from bats, and there is an increased probability that this will occur in China," Shi wrote in a review of the research in 2019. "Therefore, the investigation of bat coronaviruses becomes an urgent issue for the detection of early warning signs."[14] Every disaster movie starts with someone ignoring a scientist. It is now too late for warnings— at least for Covid-19. The warnings still hold, however, for the other bat coronaviruses that could yet invade humans.

Perhaps the team's saddest paper came out on January 29, 2020, the same day my *New Scientist* piece headlined that the new coronavirus looked set to go pandemic. This time, all the authors were Chinese scientists, most in Wuhan, and the new disease was raging in their town. They recapped the work to date: they had discovered SARS-like coronaviruses in their natural reservoir, bats. "Previous studies have indicated that some of those bat SARS-CoVs have the potential to infect humans," they recalled. The news this time was that the new virus killing people in Wuhan was 96 percent identical to one of those bat viruses, RaTG13, and used the same cellular receptor, ACE2.

We warned you, in other words. But good scientists that they were, they got on with what to do now. "Future research should be focused on active surveillance of these viruses," they wrote, and broad-spectrum drugs and vaccines should be developed against this group of viruses in general. "Most importantly, strict regulations against the domestication and consumption of wildlife should be implemented."[15]

\*  \*  \*

That last comment gets us to one of the most important questions raised by all this. Bats live everywhere. Why have these bat viruses broken out in humans, twice, in China? Is it the bats? Or the way people interact with bats?

It is actually hard to catch viruses directly from bats. Only 6 of the 218 people tested who lived near the bat caves in Yunnan had antibodies from infection with bat coronaviruses, even though they all regularly saw the bats near their homes. Similarly, MERS has been found in Saudi bats, but humans only ever catch it from camels, which apparently carry the bat virus with no ill effects. As we saw earlier, poor little Emile Ouamono in Meliandou, Guinea, caught Ebola from a bat and died, triggering the Ebola epidemic in West Africa in 2014. But children in his village routinely caught, roasted, and ate those bats with no apparent problem, says wildlife virologist Fabian Leendertz, who led the expedition to Meliandou to try to figure out what happened. He doesn't know why Emile was unlucky.

John Mackenzie of Curtin University in Australia (no relation to the author) told me that no one there has ever caught Hendra virus directly from a bat—only from horses, which might get it by eating the fibrous remnants of fruit the bats spit out or the afterbirths shed from birthing roosts. Catching Nipah requires either a pig intermediary or sharing fruit or a drink of palm sap with a fruit bat. Australian wildlife activists regularly nurse injured bats back to health, and only two, says Mackenzie, ever got Australian bat lyssavirus, a virus closely related to rabies carried by Old World bats. Sadly, they died; nowadays, bat rescuers are vaccinated.

Confused rock stars aside, people have gotten rabies by handling bats in the Americas—there's around one case a year in the United States. Most rabies in people, however, comes from ground-dwelling animals like dogs or raccoons. Britain, much of the EU, Japan, Australia, and other countries are officially classed as rabies free by the World Organisation for Animal Health, under rules governing the international transport of animals, because their ground-dwelling animals are free of the virus. Yet their bats carry lyssavirus, and the viruses and the disease they cause are effectively the same as other rabies: in 2002 a British man was infected by a bat and died of it.[16] But it is considered a far lower risk—simply because it is so rare to catch a virus from a bat.[17]

I know a woman who lives in a picturesque village in England's Cotswolds district, who rescues injured bats. She has a roomful of cages and baskets sheltering convalescent bats of nearly every species in Britain, some endangered. She freely handles them, feeds them, bandages wounds. Only one British species, Daubenton's bat, is known to carry bat rabies, she assured me on a visit, expertly lifting one out of its basket. (Since then, it's been found in another species.) They're cute little things, with incredibly soft, brown fur. I trust her judgment, but I was happy to leave handling them to the expert.

So how have SARS and Covid-19 gotten into us? Blame has fallen on the wildlife trade—especially as both diseases emerged in winter, the season for hunting and slaughtering animals in agricultural societies, so, by extension, the season when eating game meat has been traditionally considered good for one's health in China. And many of the first recognized cases were linked to a large market in central Wuhan, the Huanan market, which sold wildlife.

In April 2020, the executive secretary of the UN Convention on Biological Diversity, Elizabeth Maruma Mrema of Tanzania, called on China to shut down wildlife markets like Huanan. "The message we are getting," she said, "is if we don't take care of nature, it will take care of us."

Opinion is now divided, though, on what role that market played in Covid-19. The WHO-China study team reported in March 2021 that they could draw "no firm conclusion" about it. Of the 174 cases confirmed in a review of early cases having developed Covid in December 2019, only 55 percent had had a recent exposure to any market, and only 28 percent to the Huanan market. The earliest of these cases had no exposure to Huanan. The genetics also show this virus jumped to humans once, not many times from different animals on a market that would typically have been carrying slightly different genetic versions of the virus. It jumped somewhere, once, then spread human to human. It would not be surprising if such spread simply happened first on markets, whether or not the initial jump to humans did, the study team observed. They are, after all, places where a lot of human contact happens. That would make the market cluster a result of human interactions, not necessarily where the virus started.

Nor did it need to spread to an "intermediate species" on a market to start spreading in humans. "I am strongly of the belief that the virus as we saw it in Wuhan is pretty much exactly the virus that was in bats—it just happened to have everything it needed to spread in humans," says Rambaut. "I think the market cases were just part of a bigger cluster. It doesn't mean the market was the source." But the association with the market stood out, maybe because people associated markets and SARS. And in much of January 2020, only people with exposure to the market, or another case, could be tested for the virus—we simply don't know how many cases with no market association there were in the early weeks of the pandemic.

And just as civets proved to be, at best, incidentally related to SARS, so it seems are pangolins and Covid-19. Related viruses were found early on in the scaly creatures, the world's most trafficked endangered mammal, and Chinese scientists proposed that pangolins were the intermediate host that spread the virus to people.[18]

"The pangolin turned out to be a red herring," says Rambaut. The SARS-like viruses found in them are far less like the Covid-19 virus than are viruses found in bats.

RaTG13, the bat virus that was closest genetically to SARS-2, was 96 percent identical to it, not 100.[19] "We estimate they split from a common ancestor between 40 and 70 years ago," Rambaut told me, describing work published later in 2020. But, he says, there is no reason not to think that "the lineage that gave rise to SARS-2 was in bats for almost all of that time. I don't think we require an intermediate host to explain any of the features of the SARS-2 genome." Ralph Baric agrees, saying, "It is a mistake to assume that an intermediate host is needed."

The notion that the virus needed to have spent time in an intermediate host was also encouraged by the fact that two unusual bits of gene sequence found in the Covid-19 virus had not yet turned up in bat viruses in 2020. One, however, was found in 2021 to occur naturally in numerous bat viruses and other coronaviruses.[20] The other one did turn up in a pangolin virus, leading to theories that it evolved there. But "it's possible both of them are in a bat virus in some combination," says Rambaut. In March 2021, Chinese and Australian researchers reported a bat virus from Yunnan that was 94.5 percent identical to SARS-2 overall, but five of its genes were closer to those genes in SARS-2 than IN any other virus yet seen.[21] The variety of viruses in bats is huge, and it took fourteen years of dogged bat sampling to find all the viral gene sequences that precisely matched the 2003 SARS virus. It is not surprising that researchers haven't found dead ringers for SARS-2 or all its genres in bats yet.

Yet scientists, like all humans, have a weakness for neat little stories, even if they have been disproved. Accounts of SARS—especially by scientists who are not specialists in the precise field involved—nearly all state with assurance that civets were the virus's intermediate host. We have discussed why they probably weren't. Besides, all the SARS virus's genes have been found in bats—no need for an intermediate species. Yet the same kind of uncritical tale of Covid-19 and pangolins is now becoming entrenched in popular mythology about the virus. If that leads to further persecution of pangolins—already at severe risk due to their use in traditional Chinese medicine—it would be tragic.

So if the virus came straight from bats, but it's hard to catch viruses from bats normally, how did we get it?

Chinese scientists who study bat coronaviruses have warned repeatedly that commercial sales of bats or bat-derived materials are a threat. If Covid-19 did make that first jump on an actual wildlife market, there is hope of stopping this from happening again. Despite enormous economic costs, China shut down live wildlife markets across the country in February 2020 and issued a new wildlife law in November that banned wild animals as food, although not as medicine.[22] This may or may not stick. The lobby for wildlife sales is strong: they were supposed to be banned after SARS, but while Guangdong imposed a trade ban on wildlife meat in April 2003, it was lifted by mid-August—after SARS had disappeared—for fifty-four captive-bred species. Business as usual rapidly resumed.

Activists complain there are too many loopholes in the new law. Species farmed for fur, including mink—which were subject to a mass cull in Europe in late 2020 after they were found to be sharing a mutated strain of our Covid-19 virus—have been redefined as "special livestock" and remain legal, suggesting wildlife trade will not stop completely.[23] As well as their own germs, some wildlife can host SARS-2—so far, eighty known species can—and possibly generate dangerous new mutants.

Yet Peter Li of the University of Houston–Downtown says eating exotic wildlife is not traditional among the vast majority of Chinese. He says rural families turned to catching and then raising wild animals as a way to get food, and then to make money, after the upheavals of the 1960s in China. Since then, a wealthy and powerful industry has emerged selling ever more exotic meat to China's many wealthy city dwellers. "Demand for wild animal meat from the consumers is false," says Li. "The demand has been created by the traders and restaurant owners who claim it is good for health, longevity, sex, brain health." That suggests it may be possible to reverse the fashion, and reportedly it is less popular with younger Chinese.

It isn't just fashion, though. Bats are traditionally eaten in southern China, as well as elsewhere in Southeast Asia and in Africa—smoked bats are popular in Ghana, and Leendertz says they have become common game all over Africa as larger animals become rarer.[24] People tend to eat the meaty, large-bodied fruit bats, not tiny insect-eaters like the horseshoe bats that carry SARS-like coronaviruses—bad enough, as fruit bats have their own viral baggage. The children of Meliandou consider tiny insect-eating bats a good snack, though.

Given the risk to everyone if a contagious virus emerges in people, we should all probably stop consuming bats, as food or medicine. But it's harder than closing a few wildlife markets. Culture, poverty, and the need for protein are harder to argue with than a multimillion-yuan game meat industry—and that's hard enough.

But maybe we're on the wrong track by thinking the risk is primarily the use of bats as food. Horseshoe bats are used for TCM, traditional Chinese medicine, which is very widely used in China. The WHO reported in February 2020 that herbal medicines prescribed by TCM figured prominently in China's medical response to the Covid-19 epidemic.

Ye Ming Sha—night brightness sand—is another popular traditional medicine. It is made of dried, powdered bat feces. It's not hard to find: type it into a search engine and you'll turn up numerous online sources, although many suspended sales as news of a possible Covid connection got around. One, charging $12.38 per 100 grams, lists one of the source species as horseshoe bats. "It cools the Blood, reduces Stasis and stops pain. It treats eye disorders . . . malarial-like disorders, childhood fright, painful urinary dysfunction, vaginal discharge, scrofula and swollen sores," the product description reads. A happy customer posted, "Already put into use."[25]

But mostly, Ye Ming Sha is used for eye problems. The *Clinical Handbook of Chinese Prepared Medicines* says it "clears heats, nourishes the eyes, and improves night vision (due to high levels of vitamin A)."[26] An online site about TCM explains that "bats are blind, fly at night" so their droppings are good "for vision, especially at night."[27] (In fact, bats have excellent vision for the same reason birds do: they fly.)

Sampling in Yunnan found bat coronaviruses in horseshoe bat feces. Drying bat droppings possibly kills most of the viruses in it, but may not kill every virus, every time.[28] Demand for TCM in China is booming, partly because the government is strongly promoting it: the number of practitioners has jumped 2.5-fold since 2011.[29,30] Increased production of Ye Ming Sha may mean more virus gets through. The human eye is particularly rich in the ACE2 protein that the Covid-19 virus binds to: eyes are thought to be a major route of infection.[31,32] When I asked TCM practitioners online how to use Ye Ming Sha, they advised applying a water extract to the eye.

According to TRAFFIC, dried bodies of horseshoe bats are also a folk remedy for cough—ironic, given Covid-19's signature symptom. Perhaps the greatest risk is not from the remedies themselves, but from, and to, the generally impoverished people who catch the bats or collect droppings.[33] In 2013, Thai researchers working with PRE-DICT found betacoronaviruses, the same family as Covid, in dry bat guano collected as fertilizer and warned that collectors could be at risk.[34] A collector infected with a bat virus could spread it to other people, maybe while delivering bat products to a market—one possibility for the so-far elusive patient zero in Covid-19.

It is not unreasonable to think that China might reduce the risk of zoonotic disasters by cleaning up wet markets, whether or not those were the source of Covid-19. We know for certain that they are the source of some worrying strains of bird flu linked to live poultry sales, so China's pledge in July 2020 to phase out those sales is another promising move.[35] Peter Daszak says it would also make sense to improve biosecurity in the handling of remaining live animals on wet markets. Currently, a large number of species are stacked in cages, bats included, freely passing body fluids and any accompanying viruses among themselves. Better hygiene might reduce that.

Researchers at Colorado State University have actually set up model live animal markets inside high-containment labs to watch how viruses spread under realistic conditions. Cages of chickens, quail, pheasants, and rabbits were stacked on each other, with pigeons and sparrows flying free. Add a few birds with flu: the virus spread fast, but unexpectedly, some species transmitted some viruses, whereas others didn't, suggesting it's possible to limit the risk.[36]

But controlling wet markets is one thing. TCM is quite another. It is held in high esteem in China. Much of it is undoubtedly valuable: artemisinin, currently the world's leading antimalarial drug, was derived from a traditional herbal treatment by one of the great women of science, Tu Youyou, who won a Nobel Prize for it in 2015. All the same, some Chinese are reconsidering certain TCM ingredients. The *Chinese Pharmacopoeia* is the weighty three-volume authority for materials approved as traditional medicine in China. Its 2015 edition included bat feces, but in another promising move, pills containing it were removed from the 2020 edition—along with pangolin scales, which, press reports drily remarked, are made of the same stuff as fingernails.[37] It is not clear that taking them out of the *Pharmacopoeia* means they have been taken off the market, however—in early 2021 it still seemed possible to order Ye Ming Sha online, and there was no sign of Beijing banning it.

The larger issue, though, is conservation. If maintaining biodiversity in the wild reduces our risk of zoonosis, as scientists say, then all wildlife trade, not just China's and not just markets, needs careful consideration. Meanwhile, the use of animals for medicine in China often bypasses wildlife markets, but is driving some species to extinction, notably the pangolin. And some medicines could stem more from commercial claims or outdated theories than real, valuable traditional cures.[38]

Knowing all this about bat viruses and their transmission, one key question remains: Why bats? It seems humans can get viruses from bats, but how and why do bats get them?

For years this has been explained with a much cut-and-pasted set of proposed explanations: bats are everywhere, some live in large communities, they travel long distances. But those things are true of other species, including our own, and we don't naturally harbor things like Ebola or Covid-19. Living in large colonies facilitates the spread of bat diseases to other bats: witness the white-nose fungus severely threatening some bat species in North America. Why should that also facilitate things that kill people?

It now seems more likely to be a result of none of those factors, but instead, something unique to bat biology. Understanding this helps to explain what we have to do to stop pandemics like Covid-19 from happening again. Spoiler alert: it absolutely is not killing the bats.

Nearly a quarter of all mammal species are bats; only rodents account for more. Bats are the only mammal that truly does powered flight, using its muscles for lift rather than simply gliding like flying squirrels. In terms of evolution, this was a spectacular success.

Flying meant bats could occupy many niches—spaces in the environment that supply the shelter, food, and mates they need—that nothing else could occupy. Lots of niches meant lots of species, and two wildly different tribes evolved: the big vegetarian fruit bats of Eurasia, Africa, and the Pacific; and the little, echo-locating insect eaters of pretty much everywhere but Antarctica.

But flying has a downside: it takes a huge amount of energy. A bat's heart can beat a thousand times per minute. They burn sugars and other fuel for energy and consume oxygen just as we do when we exercise—but while flying, a little insect-eating bat does that at twice the rate of a similarly sized mouse running flat out.

All these chemical reactions generate damaged molecules, called free radicals, that are highly reactive, like fires in your cells. Bats have extremely efficient systems for snuffing these out. This has a useful side effect: long life. Free radicals are thought to cause many of the changes of aging and may be why smaller animals, with higher metabolic rates and more free radicals, live shorter lives. But whereas a mouse lives two years, a bat of the same size with an even higher metabolic rate—but a very good free radical extinguisher—can live for forty.[39]

There's another side effect, though. Bats' high energy turnover produces other molecular fragments, bits of DNA. Such fragments aren't damaging in themselves, but in us, they can mean only one thing: infection, by some pathogen that left its DNA lying around. So in humans, any such fragments trigger massive inflammation, an immune reaction that kills cells infected with viruses. However, in a bat, DNA fragments are normal, and a bat would just damage itself if it unleashed inflammation on cells that have them. So bats dial down inflammation. That means they need another way to protect themselves from infection.[40]

To do that, they evolved a different method of fighting viruses: they don't. Instead, they seem to mount a kind of nonviolent exclusion.

Cara Brook is a disease ecologist who has worked in Madagascar and California and just started her own lab at the University of Chicago. In February 2020, she published work in which she infected fruit bat cells growing in dishes with harmless viruses equipped with the exterior proteins of the lethal Ebola and Marburg viruses, both found in fruit bats. For all the cells knew, the viruses were Ebola or Marburg, and they mounted a lightning-quick response, rapidly turning on a slew of genes that effectively stopped the viruses from invading the cells.[41]

Humans use a similar response to try to stop viral entry. But we also then turn on a complex cascade of inflammation reactions to clear out cells that are infested with the virus anyway. In Brook's dishes, a few bat cells didn't react fast enough, so they were infected, the virus could replicate, and the infection smoldered on—but it was held at bay as the rapid defenses kept most of the cells virus-free. Such a low-level infection, Brook calculates, might last a bat's whole life, even if it is a relatively long one. It seems the bats tolerate a contained infection rather than risk inflammation.

STOPPING THE NEXT PANDEMIC

This could be why, even though many families of viruses live in bats, only one virus, rabies, seems to cause them much disease. Many of the symptoms of our illnesses are triggered not by what the virus is doing to us, but by the efforts of our immune system to kill the virus. That's why so many diseases, including Covid-19—and of course flu—start with the same infamous "flu-like symptoms." Given how much human disease involves misplaced inflammation—including many of the depredations of aging—it would be interesting to understand more about how bats avoid it.

But the bat viruses, like any evolving organism, fight back. The ones that attack just a little bit faster than the others are the ones that get into an occasional bat cell and replicate. Those viruses then dominate the population of viruses—or, in evolutionary terms, aggressive infection is selected for. Brook suspects this is what makes bat viruses extra deadly in us: they have evolved to beat the bat's hair-trigger response, so in us they move ferociously fast.

The mechanisms bats use to coexist with viruses could teach us a lot about how we might control our own viral infections, says Kevin Olival. And the fact that damping inflammation seems to contribute to bats' longer life spans—and might also be what stops them from getting cancer—could teach us even more. Meanwhile, apart from studying them, the best thing to do with bats is to just leave them alone.

That may seem counterintuitive. If we want to protect ourselves from viruses that normally live in bats, shouldn't we just get rid of bats? Unfortunately, people do routinely destroy bat colonies out of fear of disease, usually rabies, even though disrupting bats is more likely to spread the disease than stop it, as the uprooted bats that escape fly everywhere.

There are numerous reports of people destroying bats in misguided efforts to fight Covid-19. Colonies were burned alive in Peru in March 2020.[42] Indian states instated penalties for harming bats after mobs fearing Covid killed them during breeding season.[43] Australians shot and battered unique local fruit bats, already suffering from climate-related bushfires and heatwaves.[44] In Surakarta, Indonesia, in April 2020, hundreds of fruit bats in cages were taken from a market and burned alive.[45] In Rwanda, local authorities blasted a colony of endangered fruit bats with a high-pressure fire hose.[46] More to the point, though, "you can't get rid of bats," says Olival. "The world needs them." They are often "keystone" species on which many others in an ecosystem depend. "Bats are among the most overlooked, yet economically important, non-domesticated animals," concluded a major study in 2011.[47]

Hundreds of species of fruit, for example, depend on bats for pollination, including mangoes, bananas, avocadoes, and guavas—and if fruit doesn't rock your world, you should know that bats are the sole pollinators of the agave plants used to make tequila. The vital baobab trees of African savannas are exclusively pollinated by bats. Yet for more than half of the over 1,200 species of bat, numbers are declining—or unknown. Two species of bats in Mauritius are already extinct, yet the island nation has killed more than half of its unique fruit bats since 2015 in a misguided effort to protect fruit crops, even though bats don't do most of the damage—and nets on fruit trees work better.[48]

Insect-eating bats—like the ones that host viruses like SARS-2—can eat their weight in insects every night, and those meals include disease-carrying mosquitoes. The bats also eat tons of insects that are major crop pests. Bats are calculated to do $53 billion worth of crop protection a year in the United States alone, no polluting pesticides required—and that doesn't include the economic value of maintaining ecological balances in forests and other vital ecosystems.[49] Losing them—a real danger, scientists warn, as white-nose syndrome, rising temperatures, and wind turbines drive unprecedented bat die-offs—would trigger domino effects through ecosystems that would cost even more—and might, ironically, lead to further pandemics, as upset ecosystems and biodiversity loss are thought to be major drivers of emerging disease.

Fruit bats—like the ones thought to host the Ebola virus—are meanwhile vital for dispersing seeds in tropical rainforests. "I often say, no fruit bats, no rain forests," says Andrew Cunningham, a veteran wildlife and animal disease expert with the London Zoological Society. "In fact, given the role of rain forests in carbon storage and weather patterns, you could take this further to its logical conclusion and say no fruit bats, no humanity as we know it."

"Simply left alone, bats are harmless and highly beneficial," says Bat Conservation International.[50] Of course it would say that—but in 2006, a team of government scientists at the Arthropod-borne and Infectious Diseases Laboratory in Fort Collins, Colorado, agreed.

In a review of the research, they concluded that bats are critical for, astonishingly, nearly all biological communities on land. "Myths and misunderstandings . . . have led to efforts to extirpate bat populations, with serious consequent effects on insect control and crop production, without coincidental reduction in the already low incidence of rabies virus transmission by bats."[51]

This holds true for other viruses as well. "There has been public and political pressure in Queensland to manage Hendra virus by culling or dispersing fruit bat populations," Australian scientists reported in 2015.[52] But, they found, the amount of virus in a population of bats didn't depend on their population density, so reducing that density would not reduce the virus. Stress on the bats, though, would increase it. In 2008, researchers found that hunger made Hendra virus more prevalent in flying foxes than any other stress, making the steady loss of trees where fruit bats can live the biggest risk. And climate change and wildfires just cause more of it.[53]

The 2015 report suggested that restoring forests of native, fruiting trees to entice the bats away from people and horses would be the best way to prevent Hendra. "Bats are not the problem. They don't cause disease emergence," says Cunningham. "People do, by destroying and encroaching on their habitat and by catching, trafficking and butchering them. This can even infect other animals nearby that, if infected, might be able to carry and even multiply the bat virus, further increasing the risk."

In any case, says Olival, besides being ecologically appalling, eradicating bats would simply be impossible—there are too many, and they fly. And then the remaining bats might just carry more virus. When a cave full of bats carrying the deadly Marburg virus was smoked out in Uganda, it was rapidly recolonized by young males from other colonies, carrying more Marburg than the originals: Marburg is a childhood infection in bats. Much the same thing happens with rabies and vampire bats in Latin America.[54]

Yet Brook and her colleagues fear a burst of "misguided persecution" of bats in the wake of Covid. On one hand, they observe, people must be educated about avoiding bats and bat products so they will avoid their viruses. Yet at the same time, people must understand the "many life-enhancing services" bats provide to stop them from destroying bats for fear of those viruses.[55]

The problem with bat viruses, the researchers note, is not the bats: it is that when one of their viruses does jump to us, we let it get away. In West Africa in 2014, there was one transmission of the Ebola virus from a bat to a human child—and thousands of transmissions of Ebola between people after that. Covid-19 started with one jump of a bat virus to one or a few humans. Then that was followed by many millions of transmissions among us. That second thing—the transmission of virus between humans—is the problem.

The answer, says EcoHealth Alliance, is both surveillance, to spot and contain diseases early when they do reach humans, and conservation, to maintain intact ecosystems where bats are unlikely to encounter people or move into farms or towns, and to leave bats that do live near us alone, so the viruses don't jump in the first place. If nothing else, surveillance is cost-effective. Over ten years, says Olival, PREDICT cost around $200 million, much of it to establish ongoing capabilities to monitor emerging infections in thirty low-income countries. That is a tiny fraction of the $16 trillion the pandemic some of those labs warned of was estimated to have cost the United States alone by July 2020.[56] Top researchers estimated in 2020 that a decade of spending $22 billion to $31 billion a year on controlling the wildlife trade, protecting tropical forests, doing surveillance for new infections, and increasing the control of dangerous pathogens on farms would be only 2 percent of the cost of another pandemic like Covid-19—and that's not counting the $4 billion a year these activities would save by cutting greenhouse emissions.[57] In October 2020, the Intergovernmental Science-Policy Platform on Biodiversity and Ecosystem Services, an official scientific body that advises governments on biodiversity, reviewed all the available science, and agreed.[58]

PREDICT is itself a good example of the problems with the surveillance we have so far. The program's funding from the National Institutes of Health, the main US biomedical research funding agency, ended in 2019 and was not renewed. But it was granted another $2.26 million to continue another six months from April 1, 2020, because the labs it helped set up were, in some countries, the only ones that could detect Covid-19, and without funding, trained staff would be lost. A PREDICT lab detected the first case of Covid-19 outside China, in Thailand—the crucial trigger for China's belated recognition of the problem.

But that funding was guaranteed only through September 2020. Meanwhile, the EcoHealth Alliance's funding to work with the Wuhan Institute of Virology on coronaviruses was canceled by Donald Trump in April 2020 amid unfounded accusations that the WIV was the source of the Covid virus. Then the funding was restored—on the unprecedented condition that the US researchers who got the funding also get samples of viruses from the Wuhan lab for analysis by government scientists in the United States, and arrange access to the lab for US inspectors. There were protests from seventy-seven US Nobel laureates that this was an interference of politics in science—and, noted the EcoHealth Alliance, the conditions were impossible to meet in any case.[59] There was hope the funding would be restored without conditions under President Biden.

But this is typical of funding for emerging disease surveillance, which has been limited and, worse, capricious, dependent on varying levels of interest or capacity in the scientific world, which is focused on research, not the routine, day-in, day-out surveillance required to prevent pandemics. Even research may be hampered by increased political involvement in the science of emerging disease: for example, in April 2020, China required all scientific publications about the origins of the pandemic to clear government vetting.[60]

PREDICT has at least built countries' local capacity to continue monitoring their own viruses, says Olival: "We don't fly in, collect samples and fly out." The scientific capability they leave behind might be among the best legacies of the program. David Heymann, who headed the WHO's campaign against SARS, thinks that is what the world most needs to catch the next pandemic virus that emerges.

The more difficult question, however, is, What did we do with the warnings we got from PREDICT? The viruses they helped collect in Yunnan allowed Shi and Baric to warn us about SARS-like bat viruses that could emerge in humans with no further alterations needed. The warning was taken seriously enough that the United States initially funded PREDICT. That was about all that was done, however—and then even that became a political football as the warnings actually came true.

What about an actual response aimed at protecting ourselves from these viruses, though? The aim of the WHO's R&D roadmap is to develop vaccines, treatments, and diagnostics for its list of priority pathogens, which included coronaviruses from the start. In theory, we could have done that. In practice, as long as these viruses don't yet cause unusual levels of disease in people, there is little expenditure, whatever the WHO roadmap says. But we could at least have developed more routine efforts to watch for SARS-1 (or, as some virologists are calling it, SARS classic) and the related bat viruses in people, in case one did emerge, even resumed the drug and vaccine research abandoned in 2006. The bottom line is we didn't even do that.

Perhaps if they had known more about the bat coronaviruses the sequencing labs found in samples from patients in Wuhan in December 2019, the alarm among Chinese scientists, and the response, would have come earlier. EcoHealth and other organizations promote the concept of One Health, communication and coordinated research and monitoring of disease between researchers and clinicians who deal with human and animal health. It's a sensible idea.

But it will accomplish little as long as no one in government is tasked with the job of using this information to fund the precautionary responses it shows us we need. Or an intergovernmental forum, for that matter. We will return to that later.

It's clear: we had warnings and we didn't act on them. However, there's one disease where we *have* taken the warnings into account, for which One Health thinking and pandemic planning are well advanced: good old reliable flu.

# CHAPTER 5

# Wasn't the Pandemic Supposed to Be Flu?

> In the year of 19 and 18, God sent a mighty
> disease.
> It killed many a-thousand, on land and on
> the seas.
>
> —Blind Willie Johnson, "Jesus Is Com-
> ing Soon"

In January 2004, I went to a meeting at the venerable Royal Society in London, called to take stock of what we had learned from
the SARS nightmare that had ended six months before. The coffee
break arrived, and people from the conservation groups were chat-
ting in hushed tones about civets. It got depressing. I headed for
the coffee in the back of the room.

There I saw someone I did want to talk to. Ab Osterhaus is one of
Europe's top virologists. His lab had just "done Koch's postulates" on
the SARS virus, a rarely met standard for proving a pathogen causes a
disease.[1] And he was leaning against a pillar, looking very, very shaken.

I wasn't sure I should say anything, but Ab's a pretty informal Dutchman, so I asked if he was okay. He told me he had just been exchanging emails with colleagues in Hong Kong. "It's this H5N1 bird flu," he said. "If this adapts to humans, it could be really bad." He searched for a word. "Civilization ending," he said.

In early 2004, Ab wasn't the only flu expert who was very worried about H5N1. In fact, those experts still are. Yes, the Covid-19 pandemic is a coronavirus, not a flu. They are quite different. But we're talking here about pandemics in general. It is to be hoped that, having seen the Covid-19 pandemic, we can handle the next flu pandemic better than we might have otherwise. It would be only fair, because the last flu pandemic messed up the way we are handling Covid-19.

Flu, influenza A to be formal, is the one pandemic we know is coming. We know other diseases can go pandemic—and if anyone had any doubts, Covid-19 ended them. You can debate, perhaps, the pandemic potential of some of the viruses on the WHO's priority list. But flu is a different story. Pandemic is what flu does. You can't talk about how pandemics happen and how we respond to them without understanding flu.

First, Flu 101. Stick with me for a moment; you'll soon see why this really matters. The virus consists of eight chunks of RNA, coding for a mere eleven proteins, and a shell studded with two of those proteins, hemagglutinin and neuraminidase—thankfully abbreviated H and N. These come in different varieties, which have numbers, and, paired together, they identify the type of flu virus. Right now, two kinds of flu, H1N1 and H3N2, are circulating in humans. But flu viruses that circulate in ducks, the original host of flu, carry sixteen different kinds of H and nine kinds of N. Plus there are two more of each unique to—you guessed it—bats. Like most of the other varieties, they leave us alone.

Flu viruses adapt to particular hosts—our current two kinds of influenza A are adapted to us and don't infect birds. (There's also an influenza B, which circulates with the two influenza A viruses and makes people sick every winter, but it never seems to go pandemic, so let's mostly ignore it here.) Likewise, bird flu viruses are adapted to birds, and don't—normally—infect us. Both bird and human flu viruses can infect pigs as well as the pigs' own kinds of flu, and swap genes—and humans can catch what emerges. This makes pigs the prototype "intermediate species" in virus emergence. The flu virus is transmitted in the droplets we exhale, like Covid-19.

Droplets evaporate and fall to the ground quickly in warm weather, so flu does better in cool weather. Derek Smith of Cambridge University and colleagues worked out how this leads to the annual flu epidemic. Accidents of geography mean that in East and Southeast Asia, cool, rainy seasons are always happening somewhere, at different times. So there is always a flu season happening somewhere in the region, and flu is constantly infecting people and evolving.[2]

Then, as winter sets in in the Northern Hemisphere, the dominant flu virus in East Asia breaks out and circles the globe. Then it does the same in the Southern Hemisphere's winter. Flu basically mounts its own pandemic every year, except we don't call it that because it is routine.

The viruses that dominate this yearly tour of the planet are the ones that can dodge our immune systems best and get into the next human quickest. To get to the top of this class, flu plays a crafty game. The big H protein on its surface attracts most of your immune system's attention, and it constantly mutates, at seven different hotspots. It eventually accumulates so many little changes that many of your immune defense proteins, antibodies, that would recognize and attack the last flu you got don't quite recognize this virus. So you get sick again. The virus that is best at evading last year's immunity and infecting people dominates this year's flu.

These small changes are easy for flu: it makes a lot of mistakes when it copies its genes because it doesn't have an enzyme for fixing them. The SARS-2 virus that causes Covid-19 does have that enzyme, so it has more stable genes—although it's still an RNA virus, so it can still evolve fairly quickly. Mutation is random—and in RNA viruses, even relatively stable ones, fairly frequent, as RNA is a less stable molecule than DNA. When a random mutation happens to make a virus survive and reproduce better than others that don't have the mutation, those mutants become more successful and numerous. That is not random. That's evolution.

With flu, your immune system still recognizes the unchanged bits of the slightly evolved H and N and the rest of the virus, so you can mount some immune response and keep the infection in check. That's why most ordinary flu is just that, ordinary: just enough to make most of us miserable for a few days each winter, and—usually—no worse.

This constant change is also why we need a new flu vaccination every year. Ordinary winter flu isn't always trivial. Just like Covid-19, it is far more lethal in the elderly and people with underlying inflammation-related conditions like diabetes, so health agencies recommend a flu vaccine for such people every year. The vaccine you get next autumn needs to immunize you to the flu that will circulate the following winter—which will be a little different from the flu that circulated this past winter. But it takes six months to grow enough flu virus to make that vaccine.

So twice each year, vaccine companies and flu virologists come to a meeting at WHO headquarters in Geneva and try to predict what precise flu virus will be circulating in a bit more than six months' time, so they can start growing vaccine. They hold one meeting for the Northern Hemisphere, one for the Southern.

It's not easy. The guesswork is based on years of sophisticated observation and scientific analysis. Even so, flu sometimes pulls a surprise, and the vaccine virus the companies spent six months growing turns out to be different, immunologically, from the flu virus that ends up dominating that flu season. Or they guess right, but most vaccine virus is grown in chicken eggs, and sometimes the vaccine virus evolves and adapts to eggs, so what comes out isn't quite what was put in. In Australia in 2017, the H3N2 virus in the vaccine did this and provided little protection.[3] It's not the world's greatest vaccine technology.

There are other flu vaccines: live, weakened flu viruses you take as nose drops, and normal flu vaccines grown in cell cultures instead of eggs. But there are only a few such vaccine factories. There just isn't enough money in flu vaccines to justify much investment. Not everyone bothers getting vaccinated for a disease that in many is mild. Even if they do, it's only once a year, and the companies can't charge a very high price or they'll lose what customers they have.

A few years ago, a vaccine maker pulled the plug on a plan to build a big new flu vaccine plant in the United States: they just couldn't make it add up economically, even with substantial financial support from the US government.

Flu experts have been warning for years that we need to fix this situation, because every now and then flu makes a really big genetic change and outfits itself with an H and/or an N that few people, perhaps even no one, has ever encountered before. Then, more of the immunity to flu we got from viruses in recent years doesn't work anymore, especially if the new virus is substantially different. Such viruses therefore cause more severe disease, and we put up so little fight against them, they can circulate in seasons other than winter. We call this global epidemic of flu a pandemic.

That happened in 1918 when a particularly deadly flu virus started circulating. You will have heard about it, perhaps because of its recent 100th anniversary or perhaps because so many people are making comparisons to Covid-19. It was called the Spanish flu because it started during World War I, and news of it was censored in the countries that were fighting—but not in Spain, which wasn't. It was lethal: there were tales of people feeling slightly unwell getting on buses or trains for short trips and dying before they arrived at their destination. Opinion is divided, but fifty to a hundred million dead is the best guess—in a world that had a quarter of the population of today. In any case, it caused more deaths than the war itself.

The virus was more aggressive than most flu in attacking the deep lungs and causing pneumonia directly, and it also triggered bacterial pneumonia, which other flu viruses also do—except in

1918, there were no antibiotics for the bacteria. Historians are divided over whether the pandemic helped end World War I, but it could well have helped start World War II. Its third wave in April 1919 took the most conciliatory negotiator, US president Woodrow Wilson, out of treaty talks in Versailles, helping lead to a treaty so punishing for Germany that it is often blamed for the rise of Hitler. The United States lost 675,000 people to flu in the 1918 pandemic, more than were killed in World War I, World War II, the Korean War, and the Vietnam War combined.

Incredibly, we have actually been able to analyze that virus's genes. Virologists recovered it from an Inuit woman who had died of the flu and been buried in permafrost, and they reconstructed it in 2005. There is still some disagreement about where that virus came from and where it first broke out, but it seems to have been a bird flu that managed to adapt to humans.

Some think it got some genes from earlier human flu. If two flu viruses invade the same cell, their eight RNA bits replicate and then reassemble in random mixes. If a bird flu and a human flu invade the same cell, some of the viruses that emerge will make some bird flu proteins. These will be totally new in us, meaning our immunity to previous flu won't work as well. But it may keep some of the other genes of ordinary human flu, helping the virus spread and replicate in us.

History records what modern-day experts think were probably flu pandemics back to 1510. The one in 1918 seems to be the most lethal on record. But by 1921, that same virus was ordinary winter flu, not because it mutated massively, but because most people had encountered it, survived, and developed some immunity:

it was no longer a novel and therefore pandemic flu. It proceeded to circulate every winter until 1957, when it swapped its H and N for replacements from a bird virus—which virologists named H2 and N2, because this was all news to them and these were only the second Hs and Ns they'd seen. In retrospect, the descendant of the 1918 virus was dubbed H1N1.

That 1957 pandemic was called the Asian flu, and it killed about two million to four million people—a lot compared with the 250,000 to 500,000 thought to die worldwide in a normal flu year. In 1968, that virus swapped its H2 for what we (of course) called H3, also from a bird—"only" a million died in the pandemic of what was called the Hong Kong flu, as the change in the virus was not dramatic enough to completely defeat our existing immunity. Both viruses are thought to have evolved in southern China, which agrees with Derek Smith's findings about flu coming from East Asia.

Meanwhile, back in 1918, the pandemic virus also killed many pigs, but then they developed immunity like we did. It, too, kept circulating as agriculture modernized and hog herds mushroomed in size. Then in 1998, swine flu picked up genes from flu viruses normally found in people and in birds, both of which, as we mentioned, can also infect pigs—and hybridize with the pig viruses.

That aggressive new "triple reassortant" kind of flu virus dominated North American pig farms within a year. In 2004, virologists warned that these viruses had pandemic potential because they were also infecting occasional farmworkers and sometimes picked up H and N proteins humans weren't used to.[4]

By this time, there were two ordinary winter flu viruses circulating in people: the H3N2 virus that had taken over in the pandemic in 1968, and a descendant of the 1918 H1N1 virus, which we think escaped from trials of an experimental live vaccine that somehow reverted to the virulent strain in Russia or China 1977.[5] H1N1 had disappeared from people with the Asian flu pandemic of 1958, but a small outbreak in people of a swine H1N1 in the US in 1976 caused fears that the Spanish flu might return. Botched vaccination against it in the US fed vaccine fears for a generation, while live vaccine trials in Russia and China actually seem to have brought H1N1 back. It was fairly mild, as everyone born before 1957 had already been exposed to one like it, so the older people who are the most vulnerable to flu were safe. But younger people tended to have bad cases of flu in 1977.[6]

Then on April 21, 2009, the CDC in the United States reported two children in California with H1N1 flu—except this wasn't the 1977 strain. This was more like the H1N1 virus found in pigs, except the children hadn't been near any pigs. Then Canada issued a travel warning about a flu outbreak in Mexico—unusual in April—that Mexico hadn't reported, but that had already killed at least sixty people. The United States found two more kids with this "swine flu" in Texas. It all made ProMED. The morning of April 24, I emailed my editor at *New Scientist*: "This is exactly what an emerging pandemic would look like."

It was. Five days later, the WHO declared a pandemic was imminent.[7] We knew this because a new virus was spreading human-to-human in North America that no one seemed likely to have much immunity to. The US CDC was conducting daily briefings for us health journalists and—after a few pointed questions— finally, reluctantly, said as much. That meant it was only a matter of time before the virus was spreading rapidly on another continent.

When that happened, the WHO would officially declare a full-on pandemic. There are no hard and fast definitions for when other diseases, like Covid-19, can officially be called pandemic, but in 2009, that was the definition for a flu pandemic.

But unlike Covid-19, we knew about flu pandemics. Once it was declared, vaccine makers would activate the contracts for pandemic vaccine with the fifteen countries that had them. Countries that had pandemic plans would activate them, closing schools and handing out the antiviral drugs we have for flu, depending on how virulent it was.

I started digging. The Mexican outbreak had started in early April, killed dozens of people, including children, and spread widely over Easter—like Lunar New Year in China, the time Mexicans visit family. It started on a huge hog farm in Veracruz owned by US giant Smithfield Farms. The company protested that its pigs were all vaccinated and, anyway, had no flu symptoms. Well, yes, of course they didn't: vaccinated pigs don't show symptoms, but they can still carry and transmit flu. The vaccine doesn't stop transmission.

The UN Food and Agriculture Organization announced it would mobilize its experts "to protect the pig sector from the novel H1N1 virus by confirming there is no direct link to pigs."[8] Knowing the answer before they started must have made it an easy investigation, but of course the virus's "triple reassortant" genetic sequence made it perfectly clear that it was a pig virus. The industry mostly seemed concerned about the bad PR of people calling it swine flu. Despite their efforts, most people still do. As far as I have been able to find out, the industry still publishes no monitoring data on flu in its herds. If you were tempted to think only China is economical with the truth about pandemic outbreaks, well, it isn't.

The virologists I talked to were scared. This was H1N1, not just the same family as the H1N1 pandemic virus of 1918 but a direct descendant, passed down from the pigs we infected way back then. It also had some genes that came directly from bird flu, another unsettling similarity to its pandemic ancestor. So far, it seemed to cause fairly mild disease, although some people were dying, and they seemed unusually young. But then again, in the 1918 pandemic's first spring wave, the disease was also quite mild. In the autumn, it turned lethal. Everyone was scared that would happen this time.

The WHO was supposed to declare a pandemic once this new flu started spreading "in the community" outside North America— meaning, as it did for Covid-19, that there were people whose infection couldn't be traced to people or places already known to have it. But when Japan had a rash of cases, the WHO didn't act. Europe should have been next. Yet for some reason, cases were slow to materialize there.

On May 20, I reported why. The European CDC had set rules requiring someone to have had exposure to the United States, Mexico, or a known case to be tested, which basically precluded finding any community-acquired cases. Similar rules kept Wuhan from finding community-acquired cases of Covid-19 in January 2020, and then the UK, the United States, and other countries from finding them that February. See previous comment about being economical with pandemic truths.

The next week, two Greek students at universities in Edinburgh went to end-of-term parties, then developed a fever and a cough on the flight back to Greece. Their doctors bravely defied the rule and tested. They had swine flu. The doctors complained that the rules were stopping Europe from finding local cases. I reported all that in *New Scientist* on Friday, May 29. The rules were changed Wednesday, June 3.

I honestly don't know for sure if our article made a difference—the Greek doctors were the heroes. But after the pandemic was over, I got an unexpected gift from a senior flu guy at a health agency that shall be nameless: a commemorative swine flu T-shirt he'd had made for his staff, designed like a rock and roll tour shirt, with the dates the virus arrived in different countries. It remains a treasured possession. On June 11, with cases now mounting in Europe, the WHO finally declared swine flu a pandemic.[9]

On the day the WHO declared Covid-19 a pandemic in 2020, the agency complained about countries that still wouldn't test people who had no contact with a known case, China, or another early place that had the disease, even though it was clear the virus was spreading far more widely. It seems little has changed.

But in 2009, the WHO stressed that its declaration of a pandemic "was a reflection of the spread of the new H1N1 virus, not the severity of illness." It turned out Europe, Japan, and the United States had been begging WHO not to declare a pandemic. All their pandemic plans were based on a worst-case scenario: a pandemic of some nasty version of the bird flu that had been spreading around the world since 2004. Swine flu just didn't seem severe enough to warrant the upset. Declaring a pandemic would cause panic, they feared, for a virus that seemed to kill about as often as ordinary flu, albeit younger people.

Yet while governments were complaining that the pandemic virus was too mild to make a fuss over, vaccine makers had actually been counting on a mild first wave. That would give them time to make a vaccine in time for an autumn wave, which, if this was like the 1918 H1N1, could be dire. In the end, no vaccine was available until the autumn wave in North America was practically over anyway. There seemed to be a real mismatch between the pandemic plans, what we know flu does, and what we could realistically do about it. There still is.

Doctors I spoke to at the time said the new flu was mostly mild, except when it wasn't, whereupon it was horrendous—much like Covid-19. I remember a doctor in Winnipeg almost weeping on the phone as he described wards full of desperately ill young adults, many of them First Nations people who were especially at risk, needing ventilators and artificial breathing machines. As with Covid, death was especially targeting racial minorities and the poor.

Otherwise, it was often hard to see what made the flu almost symptomless in some people and lethal in others. Scientists at Imperial College in London actually made some good come of this by doing demanding clinical research "in the teeth of the pandemic," as one described it, taking gene samples from mild and severe cases, and discovering a gene that predisposes some people to severe disease. It's amazing how many diseases of pandemic concern are like that: not so bad, except when they are. Being able to predict who is especially at risk would help protect those people and maybe tell us exactly how these viruses turn lethal, so we can design better treatments. The researchers used that experience to get similar research underway much more quickly with Covid-19 and found five genes associated with more severe outcomes— which tells us about how the disease works and suggested new uses for existing drugs to treat severe Covid.[10]

Back in 2009, however, swine flu pulled a real surprise that shows just how different flu is from Covid-19. People born before the H2N2 pandemic of 1957, when H1N1 still ruled the earth, were more immune than had been initially expected to the H1N1 virus of 2009. You have the strongest immunity to the first kind of flu you encounter as a child, for reasons no one yet understands. Before 1957, the only human flu around was an H1N1 descended directly from 1918. So was the pandemic virus, but immunologists initially thought the two were too different, and immunity to the old virus wouldn't protect people from the new one. They were wrong.

So the old people who normally die in droves from flu—and are currently dying of Covid-19—didn't in the 2009 pandemic, one reason it's considered "mild." Something similar happened in 1918, says Jeff Taubenberger of the US National Institutes of Health: elderly people born around 1850 were also relatively immune to the Spanish flu, possibly because a flu with similar surface proteins circulated then.

You hear stories that younger adults died in 1918, but older ones did not, because in both cases their immune reactions killed them, but younger people have stronger immune reactions. Nonsense. Older people had experienced that virus before, because flu varieties come and go with successive pandemics. That happened again in 2009. If H2 ever comes back, only people born between its arrival in 1957 and disappearance in 1968 will have much immunity. I hope that's good news for some of you.

Fortunately, the swine flu of 2009 never did evolve to become more severe, possibly because it was, well, swine flu—already adapted to mammals like us and not, like the 1918 virus, largely adapted to birds and in need of some changes before it adapted to people. After 2010,

145

it settled down as a normal winter flu, like 1918 had before it. Unlike the 1918 virus, the 2009 swine flu didn't displace all the flu viruses that had already been circulating in people, only the other H1N1 that had emerged in 1977. H3N2 persisted, and it and the 2009 H1N1 now vie for dominance each winter, one or the other winning out in different places. Where H3N2 wins, more older people die.

Make no mistake: in 2009 swine flu was not benign. Three times more children died than in a normal flu season. "I think it would be very misleading to describe that as mild," said Tom Frieden, then head of the US CDC.[11] Estimates vary, but at least two hundred thousand people, and possibly as many as six hundred thousand, died worldwide from swine flu in its first year, some 80 percent of them under age sixty-five. Normally, the US CDC reports, 80 percent of people who die of flu are over sixty-five. We may not have lost many more lives than in a normal winter flu, but we certainly lost many younger lives.

All of this experience affected how we handled Covid-19 and will handle any future pandemic, flu or not. After the autumn wave, the WHO was bitterly attacked amid widespread claims that responding to the 2009 pandemic—even when we had every reason to fear a replay of 1918—was an expensive overreaction. Some now even put a positive spin on that claim: in March, Zeng Guang, a top epidemiologist at the China CDC, told the government newspaper *Global Times* that China "overreacted" in 2009, which "served as a public mobilization drill for all-round control and prevention in the face of a massive outbreak" like Covid-19.[12]

Part of the attack on the WHO was led by the kind of people who have come to be called denialists: people who reject scientific information—even observable reality—that doesn't fit with their preferred narrative, often involving claims that we are all victims of a giant conspiracy between big companies, corrupt governments, and (to them) shadowy scientists and international agencies. Swine flu was not really a pandemic, they claimed, even though it met every definition of one, and it didn't kill enough to be worth responding to. Some of the accusers seemed disappointed it hadn't killed more.

The WHO only declared a pandemic, the denialists alleged, so its cronies in the drug companies could make money selling pandemic drugs and vaccines—even though some vaccine companies actually took a loss on them, especially when a few countries, influenced (or run) by denialists, asked for their money back. Individual scientists were accused of supporting the pandemic declaration because they were in the pay of these companies—a claim that did not withstand examination but was made easier to believe because drug and vaccine companies, unsurprisingly, fund a lot of drug and vaccine research. If any of it had been remotely true, it would have made a great story for a journalist like me. It was frankly worse than untrue. It was poisonous claptrap, of the kind that has only gotten louder in the years since and has mushroomed in the info-demic of fake news surrounding Covid-19.

For example, when the Covid pandemic started, I heard people I had thought were sensible claim that this was just another scam to make money selling vaccines. (Does any reader who has had the disease believe it was a scam? And when it started, we didn't know if vaccines were even possible.) Worse, though, in the years following the swine flu pandemic, the WHO seemed to become very

147

gun-shy about calling pandemics. It stopped trying to set an official definition for pandemics at all, even of flu, so its hands wouldn't be tied in the future. In the early days of Covid-19, journalists kept asking, Is it a pandemic yet? WHO spokespersons got very annoyed. They raised the old bugbear, fear of panicking people. They asked, Why do you care so much about that word?

I will stick my neck out here and respond: because for years, the WHO has very correctly been warning us about the dangers of pandemics, especially of flu, which we know happen regularly. The word means something. The day after the WHO declared Covid-19 a pandemic, the world's press, much of which had been covering the story on inside pages, moved it to the front page. Some countries called their top-level emergency committees to discuss Covid-19 for the first time. There was an explosion of comment on social media. It made a big difference in how seriously people were taking the disease, and given the severe price we paid in many countries for reacting too slowly, I think we could have used that a week or two earlier.

Many believe the WHO is still smarting from being attacked for declaring a flu pandemic in 2009—even though it was a textbook flu pandemic—and wants to be very careful with the word. If so, the world's reaction to the swine flu pandemic did us all a disservice when Covid-19 arrived.

The WHO was also worried, however, that governments would somehow conflate the word *pandemic* with *flu*. That's one reason flu matters so much to the story of Covid-19.

When Covid-19 hit, most governments with pandemic plans had based them around flu: many, such as those of the United

States and the UK, are actually titled "influenza pandemic strategy."[13] Covid-19 is not flu, and that caused problems. Containment, where you isolate cases and trace and quarantine their contacts, was the WHO's main recommendation for Covid-19 early in the pandemic, and countries that took that advice, such as Singapore and South Korea, proved it worked. But that is not possible with flu because that virus spreads faster than Covid-19, so it wasn't a part of these official pandemic plans, one reason why several countries didn't seriously try to do containment until it was too late. The lesson here: plan, but be prepared for what we know disease can do. It isn't always flu.

To be fair, there was a very good reason governments had pandemic plans based on worst-case scenarios and flu. In 1997, eighteen people in Hong Kong were infected with a bird flu called H5N1. Six died. Virologists were shocked: it was the first time anyone had seen a bird flu directly infect people, and the results were apparently lethal. Hong Kong killed all 1.4 million chickens, ducks, and geese in the territory to stamp it out.

H5N1 reappeared in 2001 in Hong Kong—which killed all its poultry again—and 2002. In 2003, a family of four from Hong Kong contracted the virus while visiting mainland China. Two died. In January 2004, the virus ripped through poultry in Vietnam, and ten people had died of it when I ran into Ab Osterhaus looking shellshocked at the Royal Society.

He had good reason. Overall, nearly two-thirds of people who have caught this virus have died. Repeated efforts to find a large number of milder infections that would mean the real death rate was lower have all failed. It hasn't been able to spread person to person, but Ab was worried it would learn to do that without becoming appreciably less lethal. One scenario: if someone helping dispose of the millions of sick birds had human flu and got H5N1 as well, a recombinant of the two might emerge with an H and/or an N no human had seen before—even the N1 protein was different from its cousins on human H1N1 flu viruses. Add a few more bird genes, and it might not be much less lethal than the bird virus. The real nightmare might be if the bird virus itself could adapt to spread in people without mingling with a human flu, retaining the same lethality.

By late January, South Korea, Japan, and Cambodia had millions of birds sick with H5N1, and Thailand and Indonesia admitted that poultry die-offs they had been blaming on other diseases since the previous year were actually H5N1. Thailand had sick people. No one had ever seen bird flu over such a wide area. China, however, reported a few dead birds just over the border from Vietnam and claimed it had only just got the virus.

Scientists I talked to didn't believe it. In 1999, a goose from mainland China carrying the same H5 as the 1997 virus turned up in Hong Kong, and the paper reporting it was titled "Continued Circulation in China of Highly Pathogenic Avian Influenza Viruses."[14] In 2002, scientists at Hong Kong University reported finding a large variety of H5N1 viruses in chickens, which were therefore probably "now widespread in the region"—including China, the source of much of Hong Kong's chicken—and "justify renewed pandemic concern."[15] So we had reason to think H5N1 was circulating in Chinese poultry well beyond the Vietnamese border.

It turned out that after Hong Kong slaughtered all its chickens in

1997, Chinese poultry producers selling to Hong Kong had started vaccinating their birds against H5N1. That sounds like a good idea. But, American scientists told me, Mexican poultry producers had tried that, too, and discovered that the bird flu virus could circulate at low levels in vaccinated chickens, without giving itself away by causing symptoms. You didn't know it was there until an infected bird reached an unvaccinated flock somewhere—like Vietnam, Thailand, or Indonesia.

So I called the WHO official in charge of flu. I reached him on his cell phone as he sat on a bus heading up to the ski slopes—it was that time of year in Switzerland. He told me the WHO knew of virus samples from early 2003 that precisely matched the current outbreak—meaning this had been going on for at least a year. He wouldn't tell me which country they came from. On January 28, I wrote in *New Scientist* that this outbreak had started a year ago and, based on what the scientists had been saying all along, probably in China—but poultry vaccination let the virus spread unseen.[16]

The following day, China's vice minister for agriculture called a press conference in response. "It is purely a guess, a groundless guess," he fumed. "We have had strict surveillance."[17] A spokesperson for the foreign ministry said the article was "completely inaccurate, without proof and moreover does not respect science." I started getting abusive emails from Chinese students. One accused me, presuming I was British, of complicity in the Opium War.

But oddly enough, the day after the press conference, Chinese officials confirmed that there were outbreaks of H5N1 in chickens in Hubei and Hunan, quite a bit north of the outbreaks near the Vietnamese border. Two days later, there were "suspected" outbreaks in three more nearby provinces. The day after that, four more provinces had outbreaks, plus the huge western expanse of Xinjiang. Two days later, two more, this time in northern provinces.

It was as if H5N1 was, very rapidly, moving across China from its supposed toehold near Vietnam—except there is no way the virus could really have been spreading into such huge expanses of new territory that fast, even if chickens could fly. I have it on good authority that our report in fact prompted the wave, not of infection, but of candor. After all, if we could work it out, others could.

Sure enough, on February 2, *The Times* of London reported the following: "A large number of poultry markets in southern China have reported cases of the disease, and dozens of traders and butchers in contact with infected chickens have died"—but Chinese journalists had been forbidden to report the deaths.[18] There were no irate Chinese press conferences this time. Two months later, when I wrote about the further perils of chicken vaccination for *New Scientist*, Chinese state media reported it approvingly.

By then, scientists comparing H5N1 across East Asia had found it was all very closely related, but its surface proteins were changing fast. "We have a bucket of evolution going on," New Zealander Richard Webby, a leading flu virologist based in Memphis, Tennessee, told me. "This shows that H5 is circulating fairly widely somewhere, under some kind of unusual selective pressure."

Scientists who developed flu vaccines for poultry had warned about this in 2003. The vaccine might increase the risk of a flu pandemic in people, they feared, because the vaccinated birds spread the virus silently—and were a novel environment for a flu virus, so the viruses would probably evolve rapidly to adapt to it. And that might result in viruses that posed a threat to people.

The fact that flu was in chickens at all wasn't normal. Flu evolved to live benignly in the guts of waterfowl—they poop it out, other ducks drink it, the virus continues. The virus needs the duck to keep shedding it for a while until, despite being diluted in pond water, it manages to reach another duck. The viruses that didn't make ducks sick therefore won the evolutionary race.

Chickens are a different matter. Most of the nineteen billion chickens alive in the world at any one time live in large henhouses. In such pastures of plenty, a bird flu virus left by a passing duck infects chickens, then after a while often develops a "highly pathogenic" mutation in the H protein that allows it to infect cells throughout the bird, not just the gut. The virus doesn't need to persist in its host and be shed for a while before it gets a rare chance to infect another host. Hosts are everywhere. The virus that wins is the one that replicates massively and then gets into the next chicken faster than the next virus. Chickens with "highly pathogenic" bird flu die in droves, but sometimes it's adaptive for a virus to be more lethal.

In 2004, the chicken industry in East Asia, as in much of the world, had become large-scale and intensive, as growing prosperity in the wake of globalizing trade boosted the demand for animal protein. And the H5N1 virus spreading across East Asia was a highly pathogenic strain.

Usually, these viruses kill the chickens so fast they run out of them, and the virus burns out. But this H5N1 persisted because the vaccinated chickens didn't die. It did, however, have to contend with the chicken's novel immune system, which meant new pressures to evolve.

By 2006, Guan Yi of Shantou University had collected enough

anal swabs from poultry across Southeast China to demonstrate that the virus had been circulating continuously there for a decade, virtually entirely within the poultry trade.[19] There were growing concerns that it might adapt to spread readily between mammals, especially after the virus caused high-profile tiger deaths in zoos. The tigers weren't the only mammals dying of the virus. "Javanese farmers even have a word for the cat disease," said Ab Osterhaus. In theory, every infected mammal was a chance for the virus to adapt to us.

Epidemiologists, including Neil Ferguson's team at Imperial College London that did many early analyses of Covid-19, started making contingency plans for if that happened. Plan A was to watch for the first cluster of human cases and contain everyone exposed until the virus died out. If we failed to contain it—and some epidemiologists thought that was a long shot—then Plan B was to protect everyone with a vaccine or drugs. Sound familiar?

At the time, it seemed that governments did not understand that these were the only options on offer. Few had plans for deploying antiviral drugs or early surveillance to contain outbreaks—although to its credit, at this time China developed its national computerized early warning system, designed to spot emerging clusters of either resurgent SARS or newly contagious bird flu.

Plan B required developing a vaccine and, just as important, the ability to make enough vaccine and drugs for everyone. Vaccine companies have done some H5N1 vaccine development, but there was no way to make that much vaccine fast enough to reach everyone before deaths started climbing, if we started after a pandemic started. The threat is still out there, and we still can't do this for flu. We also couldn't for Covid-19, but Covid-19 had the excuse of being new. Flu is not.

In 2005, H5N1 spread beyond China, killing thousands of migrating birds at Qinghai Lake in the west of the country. Guan Yi found it was the virus from southeastern China.[20] Senior Chinese officials attacked his science and made it illegal to collect animal disease samples.[21]

I had been dubious of early reports that wild birds could carry the virus. But it started to become clear that, although this H5N1 killed many kinds of birds—diving ducks, for instance, and swans— dabbling ducks seemed to carry it with few or no ill effects, even though the virus that spread back into wild birds from infected chickens still carried the highly pathogenic mutation, the first time this had ever been seen.

That was a problem, given the wandering habits of dabbling ducks. Mallards migrate vast distances to the north in summer, nest on the tundra, and then fly back south. And the males see the world: one year they may winter in Europe, the next Africa. Between the two seasons, they dabble in the same Siberian ponds as ducks from China. I ordered something I never imagined I'd need: an atlas of duck migration.

And it showed exactly where H5N1 went. Throughout 2006, the headlines came thick and fast as H5N1 turned up in countries west of Qinghai, all descendants of the virus at the lake. There was panic in countries like Britain and Bulgaria and Germany as H5N1 appeared, killing swans here, infecting ducks there. It settled in Egypt, becoming endemic along the Nile, a beacon for migrating ducks. It turned up in northern Nigeria, just where my atlas marked a string of wetlands popular with mallards just in from Siberia. I was the target of vicious emails and blog posts from birdwatchers terrified this would mean the persecution of birds. It didn't. But the virus itself has tragically harmed wild bird populations, no one knows how much.

H5N1 is still out there, and it now has relatives, with wild birds across Eurasia and Africa carrying flu with the highly pathogenic H5, the descendant of the original H5 that started jumping to people in 1997, in combination with a range of N proteins. In 2016, H5N6 started infecting people in China, with thirty-one known cases as of early 2021, sixteen fatal.[22] All were infected by birds— like its cousin H5N1, it has not learned to spread person to person. Also in 2016, H5N8 started causing the largest outbreaks of highly pathogenic bird flu ever seen in Western Europe, killing many species of wild birds and triggering massive culls of poultry in 2020. It was thought to ignore people, but in 2020 Russia reported H5N8 infection in seven people who worked on a poultry farm with an outbreak.[23] All were mildly ill or asymptomatic.

But European researchers report that H5N8 has been hybridizing widely with local bird flu strains.[24] If any of these H5-carrying viruses swapped some of their other genes with a human flu strain, say in a pig, or acquired mutations that make them transmissible in mammals, the resulting H5 flu could pose a severe pandemic threat, as no one has any immunity to its main surface protein.

Human deaths from H5N1 diminished sharply after 2006, as people learned to avoid infected poultry. So far, seventeen countries have reported 861 human cases. More than half died, a frightening rate. Among the virus's often overlooked impacts is the cost, often to poor farmers, of destroying millions of birds to extinguish outbreaks. In Southeast Asia, that had run into billions of dollars just by 2005.[25]

H5 is not the only problem. In 2013, an H7N9 turned up in chickens on live poultry markets in China and caused severe disease in people. But until 2017, it did not have the highly pathogenic mutation, so it didn't give itself away by killing unusual numbers of poultry. Since 2013, it is known to have infected 1,568 people in China and killed 616, or 39 percent, though only four cases have been reported since late 2017.[26] This may have been because China started widespread vaccination of chickens for it that year, reducing the amount of virus people encounter—but making its spread in poultry largely invisible.

The worrying thing about H7N9 is that it appears to be partially adapted to mammals, and a few human infections with H7N9 seemed to spread from one person to another. H5N1 has been circulating widely for seventeen years now, yet it has never really done that. Virologists wanted to know if it could even evolve that capability. So Ron Fouchier—the Dutch scientist who later first isolated MERS—primed an H5N1 virus with three mutations known to adapt bird flu to mammals. Two of them were discovered in the bird-derived proteins on the pandemic viruses in 1918, 1957, and 1968, so they had a history of enabling pandemics.

Next, he infected ferrets with the primed virus in a very-high-containment lab, watched them get sick, then passed their virus to another batch of ferrets. After ten such "passages," he caged the infected ferrets next to healthy ferrets—and watched. The healthy ferrets caught the virus. This was it: mammalian transmissible

H5N1. It turned out the virus acquired two further mutations in the ferrets that allowed it to spread via droplets. So H5N1 was only five mutations, in total, away from spreading in us.

The big question: Is this transmissible H5N1 just as deadly as virus straight from the bird? None of the ferrets that inhaled the airborne virus died. However, the team had previously discovered that, because our noses are very different, ferrets can sometimes inhale—with no ill effects—viruses that would kill us. You have to get past the nasal passages and blow a virus into its windpipe to get an idea of how deadly that virus would be in people. When that was done, the transmissible virus killed all the ferrets. It had become fully contagious in mammals like us, without losing any of its deadliness.[27]

At a big flu meeting in Malta in 2011, Fouchier described how he bred the virus. As I listened to his talk, scribbling furiously in my notebook, I felt that weird mix of sensations you get on a story like this, both excited and terrified. This was big. People had started accusing scientists of hyping the threat of H5N1. Maybe it couldn't become transmissible in people, they said. *But this one did*, at least in mammals. I'll never forget how somber and serious Fouchier looked when I tracked him down at the coffee break and asked him about it. The other flu scientists I asked looked scared, too.

Later, there was an almighty ruckus when the team submitted the work for publication in *Science*, a leading scientific journal. The top biosecurity committee in the United States tried to stop it from being published, arguing that a bioterrorist might use the recipe to brew a deadly pandemic. As the work had partly been funded by a US agency, they had a say in publishing it.

Fouchier replied that we needed the work to understand what risk the virus posed, especially as it was now widespread in birds across Eurasia and Africa. But as factions for and against publishing the work dug in, and the fight dragged on, Fouchier backtracked, claiming the airborne virus wasn't really so scary, as the ferrets that merely inhaled it didn't die. After all, he wanted to publish. But of course that was not the whole story. I remember the look on his face when he told me all the ferrets died, and what I saw on the faces of the world's top flu scientists in Malta. The work was ultimately published.

What did temper the scariness a bit is that not all of those three priming mutations were known to have cropped up naturally in wild H5N1, although they had in other kinds of bird flu. Maybe for some reason H5N1 can't survive those mutations and go transmissible on its own.

But here's the really scary bit: H7N9 already has three of the five mutations that made H5N1 transmit freely between Fouchier's ferrets. The fear is that if H7N9 infects the occasional mammal, it could acquire the other mutations it needs while it's there, like H5N1 did in the ferrets. It may not need much: in 2017, some H7N9 acquired the highly pathogenic mutation in chickens, and Yoshihiro Kawaoka, a flu virologist in Wisconsin, discovered that those viruses would already spread between ferrets—and killed some of them, just from being inhaled, without the virus having to be blown into their windpipes. It's the first bird flu we've found that does that.[28] That didn't attract the ruckus Fouchier's experiment did, even though those viruses, for all we know, are out there circulating in chickens.

Of course, we don't know for sure if the mutations that made

H5N1 transmissible in mammals also work for H7N9. The experiment hasn't been done. After the confrontation over publishing the H5N1 work, further work that might make nasty viruses nastier, called gain-of-function research, was banned or discouraged in the United States and Europe. Anthony Fauci, the tough-minded head of the US National Institute for Allergy and Infectious Disease (NIAID), became a popular hero in the United States for calmly presenting the science of Covid-19 at televised presidential briefings in 2020. In 2012, he resolved the dispute over Fouchier's work by saying any future such experiments first had to be assessed for their risks and benefits by experts in the agency, or it wouldn't fund them.[29]

In 2017, gain-of-function experiments were allowed, in theory, to resume.[30] There is certainly a real risk, and not just from bad actors deliberately making a bioterror germ; possibly worse is that other scientists with leaky labs might try to repeat the work. If such a virus escaped, it would no longer matter if it might have emerged in nature—we would have shot ourselves in our collective feet. I am personally inclined to believe Fouchier's lab is as safe as they get, as the Dutch inspectors are especially stringent, but I don't know about everyone else.

This issue goes beyond flu. Ralph Baric, who found that the coronaviruses Shi Zhengli found in bats in Yunnan could infect human respiratory tract cells, wanted to see what mutations would be needed to make that virus more dangerous in people, as Fouchier was trying to find out with his ferrets. However, this could have been a gain of function, so he couldn't do the experiments. In 2019, NIAID funded the EcoHealth Alliance to work with Shi

Zhengli's lab, with its top-level containment facilities, to see what changes to the external spike protein made bat viruses better at infecting human cells.[31] It presumably vetted the risks and benefits first. The process is confidential, so all we know is that the vetting committee thought understanding the threat from these viruses outweighed the risk of escape. The funding was canceled, however, after allegations by the Trump administration that Covid-19 escaped from a lab. We will look in the next chapter at the evidence that Covid is a natural, not a lab-created, virus.

The scientists argue that we need gain-of-function work to understand those viruses better, as nature is doing its own experiment. "Mother Nature is the ultimate bioterrorist" is the mantra I hear from virologists. The viruses Shi found naturally in bats were already plenty capable of infecting human cells, and this pandemic is a good example of what Mother Nature can do. After the funding was canceled, EcoHealth Alliance released a statement noting that "international collaboration with countries where viruses emerge is absolutely vital to our own public health and national security here in the USA."[32]

In fact, Mother Nature is already hard at work on the next flu pandemic, and we could do with more international collaboration on tracking it. Even though a pig mega-farm birthed the last flu pandemic, such farms have continued getting ever more enormous since then. The more pigs are uninterruptedly in one place swapping viruses, the more scope for the viruses to evolve.

In July 2020, Chinese virologists reported that a hybrid between our 2009 pandemic flu and previous pig viruses had swept to dominance of China's pigs, and that it readily infects humans and shows worrying pandemic potential.[33] In October 2020, European virologists reported a whole tribe of flu strains on Europe's pig farms, also partly derived from the 2009 virus, that show similar potential.[34]

And in 2019, the disappearance of half of China's pigs—a quarter of the world's—due to African Swine Fever led to a huge increase in the global trade in pork and live pigs, and with it the potential exchange of viruses. It has also led to more modernization in China's huge pig industry, meaning more huge herds.

A flu pandemic is coming. That's what flu does. Maybe it will be fairly mild, like in 2009—but tell that to the people who lost loved ones then, and there were plenty. Maybe it will be H7N9, packing the same 40 percent death rate it has now. Maybe it will be a total surprise, something brewing on a giant hog farm or in a backyard flock of chickens swapping viruses with the wildlife. But it's coming. Are we prepared? No. As we will see in the next chapter, we can't make flu vaccine fast enough, in large enough quantities, for a flu pandemic. And although flu is the one virus for which we have effective antiviral drugs, it isn't clear we have enough of those, either—although a few more are now being developed.

If we're not ready for the pandemic we can see coming, how can we be ready for the ones we don't?

CHAPTER 6

# So What Do We Do About Disease?

> The world needs to prepare for pandemics
> in the same serious way it prepares for
> war.
>
> —Bill Gates, from a speech he gave at the
> Massachusetts Medical Society, 2018[1]

The world was not prepared for Covid-19, and it is not prepared for pandemics generally. "In spite of all our 'alarmist' outcries in the past for better pandemic preparedness, we are now starting to prepare when the house is on fire," says Ab Osterhaus. What should we now be doing about that?

You'd have thought we wouldn't be lacking in pandemic plans: countries and experts have been talking about it ever since the world got spooked by H5N1 bird flu in 2004. Yet when Covid-19 arrived, there were disputes in many countries about whether to lock down, how to do it, whether or not containment was possible, when to lift restrictions, whose job it was to decide, and where the scientists fit in. Instead of arguing about these things ahead of time, governments were vacillating just as medical staff ran out of ventilators and PPE, and the economic impact just of our efforts to slow the virus's spread caused mass unemployment, bankruptcies, poverty, even starvation. Very few governments seemed to have widely agreed plans for what to do when a pandemic struck, and there was almost no coordination internationally—even, initially, within the European Union.

This shouldn't have been a surprise. Christopher Kirchhoff, who led the US military's mission against the 2014 Ebola epidemic, described in March 2020, as Covid took hold, how a high-level analysis of the Ebola response had concluded in 2016 that with a trickier disease—one that, unlike Ebola, spread before it caused symptoms, like Covid-19—"the response system of the United States and the international response system would risk collapse."[2]

The United States tried to improve matters. It spent $1 billion on the Global Health Security Agenda, an international effort to build detection labs and preparedness plans in developing countries, as required by the International Health Regulations; stockpiled PPE and set up networks of hospitals in the United States primed to respond to a pandemic; and created an office in the White House to plan and lead the response, the National Security Council Directorate for Global Health Security and Biodefense. All three, wrote Kirchhoff, were underfunded or shut down under the Trump administration.[3] When the Covid-19 pandemic hit, the pandemic plan written by the Obama administration was largely ignored.

But while those political problems were unique to the United States, lack of preparation and action was not. On March 11, WHO director-general Tedros Ghebreyesus finally called Covid-19 a pandemic and said he was doing it because "we are deeply concerned both by the alarming levels of spread and severity, and by the alarming levels of inaction."[4] It's tempting to wonder if there would have been less inaction if he'd called it a pandemic earlier—a concern shared by the WHO's investigative panel, chaired by former government leaders Helen Johnson and Ellen Sirleaf, a year later.[5] But countries can't really claim that excuse. Scientists, and journalists like me, had been warning of it since January.

For weeks, the world, especially the rich West, seemed locked in a slow-motion train wreck, as though countries could not believe the oncoming storm was going to reach them and were paralyzed, not knowing how to respond. There was a lot of denial: senior officials in North America and Europe were saying this might all still be contained in China when scientists suspected it was probably already worldwide—and, it turned out, those scientists were right.[6] National plans that called unambiguously for certain responses when certain kinds of events happened should have triggered more decisive action earlier. Clearly, many countries did not have them. Even where there were plans, and even if they were followed, they were mostly devised for flu, which as we have seen is different from Covid-19 in many ways. Containment doesn't work for fast-spreading flu, but as China showed early on, it works for Covid-19. The WHO delayed calling Covid-19 a pandemic partly because it feared countries would abandon containment and testing and rush straight to flu-inspired social distancing—and for some countries, it may have been right about that.

Many countries at least tried to plan for a flu pandemic. But when a milder-than-feared pandemic hit in 2009, some countries actually rolled back even that preparation. A Global Preparedness Monitoring Board (GPMB) co-chaired by Gro Harlem Brundtland, the WHO director-general during SARS, reported in 2019 that "for too long, we have allowed a cycle of panic and neglect when it comes to pandemics: we ramp up efforts when there is a serious threat, then quickly forget about them when the threat subsides. It is well past time to act."[7]

Yes, but what action do we need? There was some hope after the 2014 Ebola epidemic in West Africa almost spun out of control that the near-miss would jolt the world into doing more to prepare for major disease incidents. And it did trigger a few things that have been invaluable in dealing with Covid-19, such as the creation of CEPI, which organizes funding for pandemic vaccines, and the beefed-up emergency response capability at the WHO.

But we were still caught flat-footed. Work on drugs and vaccines for coronaviruses had been minimal, even though we knew the risk. Some basic research had been done, and some start-up companies even had a few experimental vaccines, but nothing ready for prime time. Pandemic plans varied from country to country or even state to state or didn't exist. A high-level UN panel warned in 2016 that the world was underestimating the risk of something less readily controllable than Ebola—such as a virulent respiratory pathogen—and that its ability to prepare, never mind respond, was "woefully insufficient."[8] Covid-19 is a virulent respiratory pathogen, and they were right.

The "need for speed" the WHO warned of failed to register in all but a handful of places, famously in South Korea, Taiwan, Singapore, Hong Kong, Japan, and New Zealand, and also in places that got less attention: Vietnam, Senegal, Sri Lanka, Rwanda, Iceland, and Ghana. After Ebola, the UK set up a rapid response team that, it proudly said at the time, could investigate and respond to disease outbreaks any- where in the world within forty-eight hours.[9] But when Covid-19 arrived in the UK, the response was far slower. An initial plan for quarantine and contact tracing to achieve containment was let down by inadequate testing, then it was abandoned for a scientifi- cally half-baked plan to allow most people to be exposed in order to develop "herd immunity." This was in turn abandoned when scien- tists explained the large number of deaths this would entail.

It was replaced by social distancing, as there were already too many cases to contain. But the delay, plus weak enforcement, led by May 2020 to Europe's highest death rate. Continued half-measures, repeated delays in enacting scientific advice, the failure of dubious private contractors to provide vital services like PPE and contact tracing, and a popular refusal to obey quarantine or lockdown instructions—itself possibly the result of government vacillating— kept UK case numbers high, giving the virus opportunities to evolve. A more transmissible strain emerged in England in September 2020.

Meanwhile, limited protective equipment took a high toll of health care workers in Britain and many other countries. The GPMB found that "the great majority of national health systems" couldn't handle the large influx of patients they would get with a severe, fast- spreading respiratory pathogen. There was no surge capacity in hos- pitals, they charged, or in key manufacturing, like making medical masks and gowns.[10] Covid-19 proved that and many of the GPMB's assessments correct—most horrendously, in India in April 2021, where Prime Minister Narendra Modi, like other populist leaders, had neglected pandemic control measures, and a massive surge in cases rapidly overwhelmed the country's health system.

At least the world's governments now have plenty of evidence that all those warnings about pandemics weren't overblown doom-mongering by scientists. The risk is real. That could be the real silver lining of this pandemic: there is no longer any dodging the fact that humanity is at risk from fast-spreading infectious disease, and currently many countries can apparently do little to prevent it or to respond effectively. On March 26, 2020, the G20, the world's twenty biggest economies, issued a statement promising "to strengthen national, regional, and global capacities to respond to potential infectious disease outbreaks by substantially increasing our epidemic preparedness spending."

And they did their homework on what the spending should cover. "We further commit to work together to increase research and development funding for vaccines and medicines, leverage digital technologies, and strengthen scientific international cooperation . . . rapid development, manufacturing and distribution of diagnostics, antiviral medicines, and vaccines, adhering to the objectives of efficacy, safety, equity, accessibility, and affordability. We ask the WHO . . . to assess gaps in pandemic preparedness and report to a joint meeting of Finance and Health Ministers in the coming months, with a view to establish a global initiative on pandemic preparedness and response."

And this time, no dropping the ball when the crisis is over, they promised. "This initiative will . . . act as a universal, efficient, sustained funding and coordination platform to accelerate the development and delivery of vaccines, diagnostics and treatments."[11]

It sounded good. But there was no statement launching or funding this initiative after the promised meeting took place—online—in September. Then the final summit of that year's G20 round in November again failed to commit to making this initiative a reality: it released vague statements about "advancing global pandemic preparedness, prevention, detection, and response" and "continued sharing of timely, transparent" information.[12] Under the circumstances, they might at least have pledged the "improved" sharing of timely, transparent information about disease outbreaks, given that the failure to do that, as we saw in Chapter 1, helped make Covid pandemic.

What exactly should we be working on, though? We don't know what virus will strike next, apart from flu—and we don't know what kind of flu. The March G20 list is a good start, but what is a "substantial" increase in funding from governments reeling from the multibillion-dollar cost of the pandemic? The planning effort will at least be aided by the fact that now we can see clearly what we should have been doing for the past ten years. Let's look at the toolbox.

First, know your enemy. What germs should we target? And must we focus exclusively on responding to a disease once it emerges? Can we do more to stop it from emerging at all?

We looked earlier at the WHO's list of priority pathogens to get a feel for what may be out there. But not everyone thinks such lists are helpful. In 2018, the Johns Hopkins Center for Health Security warned that lists like the WHO's "stultify thinking on pandemic pathogens" by suggesting those are the only diseases we need to worry about. And some pathogens on the WHO list, they implied, were not truly global risks but were put there to please regions where they are a problem—Lassa and Rift Valley might be examples. Instead, Hopkins called for keeping a close eye on the entire class of pathogens they suspected was most likely to cause real problems: respiratory RNA viruses.[13] They mutate and evolve faster than any other pathogens and thus can jump species fast. And, contends Amesh Adalja at Hopkins, although we can stop gut pathogens with sewage management and infections like Ebola and HIV by being careful about body fluids, respiratory infections are harder because no one can stop inhaling. Two years later, Covid-19, a respiratory RNA virus, confirmed their suspicions.

Hopkins also called for more investigation into what pathogens are out there actually making people sick. Many people don't realize that most diagnoses doctors make are "syndromic"—pneumonia, meningitis, fever, and sepsis are terms that describe the disease process, not what is causing it. The actual pathogens behind these are often not even determined, as this is not needed for treatment. Instead, doctors use wide-spectrum antibiotics for bacteria, or with viruses—like Covid-19—they just try to keep the patient alive until the patient's antiviral immune response kicks in.

"Illuminating this biological dark matter," the Hopkins team argued, "would focus pathogen discovery efforts on established damage-causing microbes." Aggressively doing such diagnosis in a few sentinel locations, perhaps in zoonosis hotspots, might reveal the next big threat early as it starts jumping to people. It was for just such a research project that Wuhan Central Hospital sent a sample of its mystery pneumonia to Shanghai, and the sequence was revealed. But that also revealed the problem: by the time Shanghai got that sample, the virus had already spread widely.

## So What Do We Do About Disease?

Once we find something new, we have to move fast. Joel Wertheim at the University of California and colleagues found in November 2020 that a novel animal virus might jump many times to individual people then get no farther, as, at first, not everyone passes it on.[14] That happened with Covid: the first cases that traveled to Europe and North America didn't pass the virus on. In fact only some cases ever do—around 80 percent of people with the virus caught it from only 20 percent of the infected population. An epidemic of such a virus doesn't take off until there are enough cases around for a few of these efficient transmitters to emerge.

And then it snowballs because when it does transmit, the SARS-2 virus spreads very efficiently—those R values of two or so for the population are an average of the many who don't pass it on, and the few who do. Then the problem is that Covid has fairly low death rates and often unremarkable symptoms, so the number of cases is too small to be noticed for some time. Then the exponential curve kicks in, and it is noticed—but it rapidly becomes too late to stop it. Hence George Gao's tears in early 2020: he realized that was what had happened.

Wertheim's team calculated that Covid had to have jumped to people in Hubei "somewhere between early and mid-October 2019" to generate the small genetic variations in virus samples from people in Wuhan in early 2020. But doctors didn't spot unusual clusters of pneumonia cases till December—whereupon, as we have seen, cases rose rapidly and might have been hard to contain even if Wuhan had started trying immediately.

So how do we spot the next one? During the pandemic, monitoring wastewater treatment plants for the SARS-2 virus revealed its local rise and fall, enough to inform public health decisions.[15] Wertheim suggests looking in archived sewage samples in Hubei to establish when SARS-2 first emerged.

But that's looking at the past. "I'm not as confident that novel human viruses with epidemic or pandemic potential could be predicted from screening waste water in real-time," Wertheim told me. It would reveal lots of viruses, but what should we make of them? "Unfortunately, the only sure-fire way to know if a pathogen is efficient at transmitting among humans is to see if it can transmit among humans." Better screening in people and communication of results—including viruses that may not initially appear threatening—is, he suspects, the only way forward.

Either approach—monitoring sewage or monitoring people— would require us to detect viruses we haven't seen before.

That's not impossible. To do it, hospital labs need new kinds of diagnostic technologies that can distinguish a wide range of pathogens. This is why "diagnostic tools" is one of the things the G20 said, in March anyway, they'd make sure we get. Fortunately, there has been a massive surge in these over the past decade.[16]

Diagnostics manufacturers make automated panels of tests, based on the same PCR technology used in the Covid tests that require a deep nasal swab. These panels can recognize the DNA or RNA of, say, a dozen respiratory or gut viruses in one sample, a huge improvement over old methods based on growing pathogens in culture to identify them, which is slow and insensitive. Making this kind of capability available more widely, including in countries that cannot afford it now but have hotspots of disease emergence, would illuminate a lot of pathogenic "dark matter" and give us a much stronger idea of exactly what pathogens we are contending with. With the vast uptake of Covid diagnostics worldwide, this could be a good impact of the pandemic.

So could the fact that Covid moved us beyond these PCR tests. One problem we encountered with Covid was that, when lockdowns were lifted, case numbers usually rebounded, as we could not tell who was infected and should remain socially distanced, and who was safe. It was all or none: lock everyone down, or not. We couldn't lift lockdown only for the uninfected, so lifting restrictions as average rates of infection went down nearly always meant they would start rising again.

Scientists have advocated for simple, cheap, and rapid "antigen" tests for the SARS-2 virus to ease such transitions and determine who can safely enter an office, go to school, or get on an airplane, and those who cannot. Although they are less sensitive than the PCR tests, the rapid tests for Covid-19 are sensitive enough to pick up who has enough virus in their nose to pose an infection risk—the only thing that matters. And they are fast enough to be used in the day-to-day management of potentially risky situations.

The advocates were opposed by old-school diagnostics developers for whom maximum sensitivity was paramount, even though the PCR tests are so sensitive that they often diagnose as infected people who merely carry the noninfectious remnants of the virus from an infection that has run its course. It is a good example of the kind of debate we should work through when a pandemic is not raging, if only because profits from massive lab-based testing during a pandemic might well cloud the issue. Certainly it seems fast tests could help us emerge from Covid. One hopes this will lead to more research on tests like this, and on how and when to use them to control pandemics. We could certainly use a more selective, less economically destructive approach to stopping people spreading a disease than wholesale lockdown.

But that's pandemic management. How about routine surveillance of the viruses circulating in people that might stop pandemics from ever happening? Our growing experience with rapid antigen tests might allow us to screen people more routinely for worrying virus families or emerging viruses of concern. Now, most widely used diagnostic panels on the market are designed for routine hospital practice and mostly look for the usual suspects that cause most human infection. In sentinel sites looking for surprises, it would be good to have something that can spot the unexpected and unknown. Moreover, the existing machines are designed to diagnose sick people already in hospitals. Wertheim's work suggests we need to catch emerging diseases earlier by routinely screening people generally for novel pathogens.

That might seem impossible—how do you design a test for the unknown? But one system, IRIDICA, put on the market in Europe in 2014, could do just that. And it showed that as with drugs or vaccines, the tough problem may not be the technology but, once again, the market for it.

The IRIDICA system was based on replicating the DNA or RNA from the pathogen in a sample and then putting it through a small mass spectrometer, which precisely determines its molecular weight down to the last atom. Using a database of known weights from different pathogens, this allows you to identify the species, and even whether bacteria are carrying an antibiotic-resistance gene. Or if it doesn't match any known species, says Rangarajan Sampath, chief scientist at FIND, a nonprofit in Geneva that promotes diagnostics development, it can tell whether it is a hitherto unknown flu, or coronavirus, or a member of other virus families.

It was initially sponsored by DARPA, the US defense research agency that (DARPA people always mention) invented the internet. It was originally intended to scan for biological weapons—I heard about the prototype at a biodefense meeting in Stockholm in 1998. In 2009, an experimental prototype of the technology was the first to spot the new swine flu from Mexico in the United States. It was set up to recognize not just flu but each of its eight RNA elements—and it spotted that this new virus had elements from bird, swine, and human flu viruses.

But as with so many promising approaches to infectious disease, this machine failed the final test: economics. Doctors and regulatory agencies for medical technologies have been slow to warm to automated diagnostics. IRIDICA was finally put on the market in Europe in 2014 and was on track for approval in the United States. Then, in 2017, the pharmaceutical giant Abbott, which owned IRIDICA, simply stopped making it. It had been a tough sell to hospitals with budgets stretched by government cuts and the growing health demands of an aging population. The real problem was that identifying the pathogens causing their patients' fever or pneumonia wasn't obviously cost-effective if there was often no pathogen-specific treatment to use as a result.

"It deeply saddens me," says Sampath, who had great hopes for IRIDICA. "There is still no viable alternative," he laments, especially for rapidly diagnosing pathogens causing sepsis, which is often lethal and where knowing the pathogen fast really can save a life. It would be invaluable in places like zoonosis hotspots, where real novelties can turn up, as it can rapidly rule out virtually all known pathogens and pinpoint an unknown virus's family.

That said, another silver lining of the Covid-19 pandemic is that it has revealed how bad we are at diagnostics. New tech may be on the way. In December 2020, the main US biomedical research funding agency, the National Institutes of Health (NIH), announced $107 million for a program called RADx-rad to find new diagnostic technologies. One involves wastewater monitoring.[17] Some seem likely to involve CRISPR, the Nobel-winning technique for precisely manipulating RNA or DNA, which has been the basis for several new Covid tests and should be tweakable for other pathogens.[18]

Then there's smelling for pathogens. We already use sniffer dogs and rats to diagnose diseases by smell; devices might soon mimic that by detecting tell-tale volatile molecules from pathogens and even immune reactions in people's breath, on skin, or just the air in crowded gathering places.

The aim generally is to make diagnostics sensitive, specific, fast, and on the spot, with no need for long detours via labs. But for now we know our quarry—Covid—and most work is focused on that. There has been less attention to technologies that can name the unknown.

There is a vicious circle in trying to promote more specific diagnosis. As noted earlier, there's no point testing infections routinely if there isn't a specific treatment for the pathogen you find. But since we don't test, the Hopkins group argues, we don't know what pathogens we should be developing treatments for. Astonishingly, when they wrote that in 2018, there were no specific treatments or vaccines for any respiratory RNA virus besides flu. At least we now have a few vaccines for Covid-19.

Moreover, there is global surveillance only for flu. Countries determine what varieties they have circulating and send samples to a global network of labs organized by the WHO, which is how we keep tabs on flu evolution and make new vaccines every year against what we think will circulate next. One perk for participating countries is that, in theory, they get access to any pandemic vaccines that result—although in an emergency, with vaccine-manufacturing countries tempted, despite international agreements, to hang onto whatever vaccine they make, it isn't certain the guarantee will work. The way some rich countries snatched up Covid-19 vaccines in 2020 despite the WHO's efforts to promote international sharing reinforces those fears.

The Hopkins team wants surveillance extended to respiratory RNA viruses other than flu, with sampling from around the world, especially in suspected hotspots. They want it to include corona-viruses, Nipah and Hendra viruses, and enteroviruses, the most common family of viral infections in humans, which are mostly symptomless or mild—except a few that, sometimes, most certainly are not: one of them is polio. They even want to watch rhi-noviruses, the only more frequent cause of common colds than the four mild coronaviruses that circulated in people before Covid-19. After all, one of those coronaviruses had a relative that went bad.

Some think we should find our plagues before they find us. Of course, we should try to spot new diseases after they emerge in people so we can shut them down fast. But Peter Daszak observes that with both novel disease outbreaks and their economic impacts increasing fast, it would make sense to deal with the underlying drivers—changes in human ecology and human–animal interactions—to stop emergence from happening at all.

To enable that, the Global Virome Project wants to genetically sequence and map the estimated half-million viruses in animals and birds that belong to families of viruses that we know can infect humans. It would cost $3.7 billion over the next ten years, says project leader Dennis Carroll, who also launched the PREDICT program that helped discover the bat viruses in Yunnan. He contrasts that to the trillions the Covid-19 pandemic will cost. Knowing where the potentially dangerous viruses are, he says, will help us focus preventive efforts, such as reducing interactions between people and the species, products, or places that we know have worrying germs. Once again, of course, that will require following up the surveillance with meaningful action.

"What we don't have is a well-coordinated global effort to monitor these viruses in their natural habitat, and prevent spillover," says Carroll—or if we can't prevent it, then detect it as it happens and not months later when it has spread. Specifically diagnosing more human infections might spot things once they've jumped, but in addition Carroll wants to learn more about the thousands of viruses that are still only in animals, to get a better idea of what might jump—and prevent it.

To do that, like its forerunner PREDICT, the GVP would set up local capabilities to monitor viruses, and just as important, establish regulatory frameworks for sharing samples and the benefits they may bring—a real problem in some countries now that won't share pathogen samples internationally out of concerns for gene "ownership." Carroll was hoping to start in 2020 with initial funding from several governments, but the pandemic, ironically, derailed that.

Critics object that although such a survey of global viruses would be great science, it isn't much good for preventing the next pandemic unless, as Wertheim noted, we also know what the viruses we discover can do in humans. "These efforts will not necessarily translate into better pandemic preparedness, given the sheer numbers of viruses that will be catalogued without a clear means of prioritizing them, [and] the fact that most identified viruses will pose little to no threat to humans," says the Hopkins team.

The GVP counters that it hopes to find markers for which viruses are likely to be virulent or contagious. But "no amount of DNA sequencing can tell us when or where the next virus outbreak will appear," argued Andrew Rambaut and colleagues in a critique of the idea published in 2018.[19] The 2014 Ebola epidemic was, at that point, the most-sequenced viral outbreak of all time—and that didn't stop another in the Democratic Republic of the Congo in 2018. Indeed, by 2013, virologists had sequenced and reported SARS-like bat viruses able to infect humans and warned of their pandemic potential. "That prediction didn't stop Covid," says Adalja. "People think the risks of these things are hypothetical." As we all know now, that one wasn't.

Echoing the Hopkins group, Rambaut and colleagues say it would be money better spent to do disease surveillance in people, to spot new infections as they emerge, using sequencing to spot the viruses and serology—testing blood for antibodies—to see what infections people have had before.

And this, they say, would be best done by a global network of trained local researchers. There Carroll agrees, although he also wants this network to "monitor, respond and prevent viral spillover while they are still evolving in animal populations."[20] (Case in point: bat viruses.) Even scientists who disagree on where in the process of viral emergence we need to look agree on one thing: we need more people, everywhere, looking, preferably in their own backyards.

We do have some eyes on the world already. The Canadian system that first spotted SARS is still watching global online chatter for mentions of disease and sends WHO about three thousand "signals" per month, online mentions of things that might bear watching. The WHO follows up on about three hundred of these and investigates thirty in more detail—on average, one a day.

But veterans of the international health scene, like David Heymann, and Seth Berkley, head of the vaccine agency GAVI, say more countries should do their own monitoring, then share the results. Eavesdropping on global online chatter is an interesting way to keep tabs on things people have already detected, but to do that detecting in the first place, local public health people with a good idea of local diseases, and the ability to investigate, would be optimal.

This is not a new idea—it just hasn't been acted on. The International Health Regulations (IHR) were originally a binding international treaty based on earlier rules, first drawn up in the nineteenth century, that required countries to notify each other about a few diseases—in the 1960s these were cholera, plague, yellow fever, and the yet-to-be-eradicated smallpox—that posed an international risk through shipping.

## So What Do We Do About Disease?

After SARS, the IHR treaty was revised. The 2005 version obliges countries to save lives and jobs endangered by the international spread of any disease. They are supposed to coordinate their monitoring and response to disease with each other, and rich countries are meant to help poor ones do sufficient surveillance to spot anything dangerous.[21]

They have done a bit, but not enough. When Ebola broke out in West Africa in 2014, the first thing that failed was surveillance. The outbreak began in Guinea in December 2013, but it wasn't identified as Ebola until March, by which time it had spread widely. After that, the response failed: it was August, and the virus was out of control in two cities before the WHO declared an emergency.

The WHO was criticized for the second delay, which happened partly due to organizational rigidities it has since tried to fix.[22] But the fundamental problem was the first delay, and that was failed surveillance. The IHR requires countries to tell the WHO about any outbreak that is serious, unusual, or could trigger international travel or trade restrictions. That applied to Covid-19, and China did tell the WHO about it, but there are no provisions allowing the WHO to inspect the situation on the ground to see if the declaration is true—for instance, whether the infection really wasn't spreading between people. Nor are there requirements for countries to share actual samples of the pathogen.

Most concerns, however, have focused on countries much poorer than China that don't have the capacity to detect and diagnose a sudden cluster of infections and tell the WHO about them. Many such countries are in exactly the tropical or subtropical hotspots for disease emergence that need closest watching—it was to investigate such alerts that the UK developed its forty-eight-hour response team. When the IHR treaty was updated in 2005, it required all countries to put their own surveillance capability in place by 2014. It then extended the deadline to 2016. Did countries make it?

"They didn't," says David Heymann, who steered the negotiations that revised the IHR in 2005. "Rich countries have been more interested in funding international response capabilities," like the WHO's new emergency unit. "There's been much less help for poor countries to take charge of their own pathogens surveillance." It's almost as if rich countries are interested in riding to the rescue in emergencies, but not in preventing the diseases that cause emergencies in the first place. In fact, surveillance and response need to go hand in hand.

An assessment in 2019 by the Global Health Security Agenda, aimed at measuring and enabling countries' adherence to the IHR, found that "no country is fully prepared for epidemics or pandemics," whether it was rich or poor. Countries were judged on, among other things, whether they could prevent the emergence of pathogens, detect—and report—epidemics "of potential international concern," and respond to them, treat the sick, and protect health care workers.

The results were abysmal. Only 19 percent of countries got marks over 80 percent for detection and reporting capabilities, and fewer than 5 percent got top marks for their ability to rapidly respond to and mitigate an epidemic. The overall average in all categories was 40 percent, rich and poor together, but even rich countries on their own got an average score of only 52 percent.[23]

## So What Do We Do About Disease?

That should come as little surprise after Covid-19. The real surprise was that what the GHSA thought was good preparation didn't work out that way. The ten countries that did worst at handling Covid-19, with the most deaths per million inhabitants—including the United States and the UK—had ranked among the top twenty in preparedness scores. The higher death rates were more than could be explained by such richer countries having, on average, older people—and besides, high scorers also took longer to find their first case. In general, the lower countries scored on the assessment, the better they did with an actual pandemic.[24]

That seems like a paradox, but it came down to something that wasn't assessed by the GHSA. Recent experience with infectious disease—knowing the need for speed—helped with Covid, witness the successful response of poorer countries that have more such diseases, such as Vietnam. Yet those countries scored lower on the preparedness assessment, as they could afford fewer of the recommended measures.

The trust of people in government and science also mattered, says Ngaire Woods of Oxford University, as did the ability of governments at all levels to pull together and cooperate with their own communities. Worst of all, she says, were "strategies aimed at political popularity rather than the pandemic."[25]

Pandemic preparedness surged a bit in rich countries after the anthrax attacks of 2001 in the United States and after the threat of H5N1 bird flu in 2004. But when the 2009 flu pandemic wasn't apocalyptic, preparedness fell out of favor. Some countries let stockpiles of antiviral drugs for flu lapse, and as far as I have been able to find out, no one renewed preorders for pandemic flu vaccines.

The UK government held a simulation drill for a fictitious H2N2 flu pandemic in 2016, Exercise Cygnus, involving nearly a thousand officials—a bit like the simulation Johns Hopkins did of a coronavirus outbreak in 2019, but involving the actual government. It refused to publish the results until October 2020, after repeated public demands and the threat of court action. It was soon clear why.

In the simulation, health services, and even morgues, were quickly overwhelmed, but when then health minister Jeremy Hunt had to authorize taking severe cases off life support to make way for patients with better chances—as Italy had to in February 2020—he flatly refused to keep playing.[26] The exercise produced recommendations for improvements in England's preparedness that would have saved thousands of lives from Covid, such as stocks of PPE for hospital workers, surge capacity, and help for old people's homes in handling discharged hospital patients, plus yet more PPE.[27] The design of Cygnus itself betrayed a confused idea of pandemic management: it contained no social distancing, the standard flu response. When queried about this in 2020, officials still seemed not to understand, sniffing that Cygnus wasn't meant to simulate "other diseases."

A similar government simulation in the United States in 2019, Crimson Contagion, also not released until it was leaked by the *New York Times* in 2020, starred a hypothetical, highly contagious H7N9 flu from China. It revealed many of the same things as Cygnus in the UK: insufficient funding for pandemic response, poor communications within and from government, confusion over which agency did what, garbled plans for schools, insufficient PPE or ventilators or capacity to make them. Both US and UK assessments were vindicated by Covid.[28]

## So What Do We Do About Disease?

You would think if any organization was prepared for epidemics and pandemics it would be the WHO, but its slow response to Ebola in 2014 proved otherwise. That was partly due to undue deference to its own African regional office and local governments, which tried to downplay the outbreak when it still seemed small, another case of wrongly regarding the economy and health as a trade-off and reacting too slowly to exponential spread. It was also, I heard at the time, stymied by a rigid hierarchy that kept epidemic experts in the field from alerting the WHO leadership to what they knew was coming—that difficult point in an epidemic when it still looks trivial, but isn't. Shamefully, it was also because member states didn't supply the cash for an emergency response fast enough once the WHO realized more was needed.[29]

But the WHO adapted fast. "During Ebola, we needed to do new things, like recruiting 2,000 people to isolate Ebola cases and their contacts in remote areas," says the WHO's Bruce Aylward, who ramped up the agency's Ebola response in September 2014. The WHO had to bring in new kinds of expertise from disaster response agencies. The learning curve was steep. "It's like asking a penguin to fly," he told me later. "You throw it off a cliff, and I'm amazed how well the damn thing flew."

The WHO emergency response team that grew out of that experience led the international response to Covid-19: the penguin has grown real wings. But it is supported by voluntary funding from member states, never completely reliable, and repeatedly imperiled, including at the height of the pandemic in 2020 when the United States left, taking a quarter of the budget with it. That has at least been put right by the Biden administration. Yet to prevent future pandemics, the world needs reliable surveillance and response, and the kind of national planning and funding Cygnus and Crimson Contagion showed were lacking. And those need to be coordinated globally: the job of the WHO. But both the WHO and national pandemic agencies will need more funding, more reliably, to do that.

The Global Preparedness Monitoring Board, which gave the world's countries such bad marks for preparedness in 2019, has more right than most to say "we told you so." In September 2020 it issued a follow-up report that stated, "Financial and political investments in preparedness have been insufficient, and we are all paying the price." It called for "sustained investment in prevention and preparedness, commensurate with the scale of a pandemic threat." Six months into the pandemic, that clearly was not happening yet.[30]

But it will take more than throwing money at the problem. What we do will have to be joined up, so at some level findings give rise to action. With more surveillance, for example, we will need to be able to predict, far better than we can now, which of the outbreaks we uncover pose a threat and require a response—then ensure the response happens. Covid-19 was not the first unexplained pneumonia in China—earlier we saw there were instances reported on ProMED in previous years that seemed to go no further, and Wertheim's team found that zoonoses can involve a lot of those. Which outbreaks have legs, and what might tip us off?

There is hope that "big data" might help, with everything from Google searches for "flu" to hospitals' anonymized electronic medical records crunched on a large scale to suggest when something serious might be emerging. Ultimately, more research into how viruses damage us, and what makes some worse than others, should start telling us how to spot the dangerous ones, as Carroll hopes. Systems like China's reporting network, designed to reveal an unexpected cluster of a syndrome in a region before it may be obvious on the ground, would certainly be valuable if it were installed in more places—provided it is used.

In January 2021 the WHO's own panel investigating its response to Covid agreed. The WHO's methods for tracking emergencies, they complained, "seem to come from an earlier analog era and need to be brought into the digital age." An information system fed by local people and real-time data gathering, they said, is needed "to enable reaction at the speed required—which is days, not weeks." Of course, the technology has to be accompanied by a step change in countries' willingness to be accountable for acting on the alerts.[31] For some countries, that could be a big step.

There are more futuristic scenarios. In March, in a report for the think tank the American Enterprise Institute, scientists called for a permanent national infectious disease forecasting center in the United States, to "function similarly to the National Weather Service" and provide "decision support" for public health, including what responses are warranted by what kinds of events. In January 2021, President Biden's new administration announced that it would set up a National Center for Epidemic Forecasting and Outbreak Analytics, as well as "modernizing global early warning and trigger systems . . . to prevent, detect, respond to, and recover from emerging biological threats."[32] During Covid, governments called in disease modelers, such as Neil Ferguson's team at Imperial College London, to predict where the epidemic was going and what effects different policies would have—the Imperial team was credited with dissuading the UK government from a lethal reliance on "herd immunity." Caitlyn Rivers at the Johns Hopkins Bloomberg School of Public Health says that happens on an ad hoc basis in epidemics, but there is no constant investment in both modeling and the public health data that models need to improve the forecasts, or any standing, authoritative agency to advise government during disease emergencies.[33] With more such agencies in more countries in the wake of Covid—and perhaps a formal, funded global collaboration among them—we might bring better science to bear on predicting and controlling outbreaks far more effectively in future.

The researchers warn, however, that predictions in any complex system, as disease outbreaks undoubtedly are, are no small undertaking.[34] Weather is a fitting analogy, as the US National Weather Service spends $1 billion a year collecting massive banks of weather data to turn into forecasts. The US CDC spends a quarter of that on analogous data for public health—local incidences of disease, age-specific death rates, vaccination rates—and has no budget for forecasting, as that has never before been realistic in public health. It's not yet clear what we need to make it realistic now.

Of course, we already have one proven, crowd-sourced, worldwide, and battle-hardened disease surveillance system: ProMED. The WHO's investigative panel gave it a backhanded compliment in January 2021, when it complained that WHO was getting more and better outbreak alerts from "platforms to collate epidemic intelligence from open and non-traditional sources" than from governments.[35]

ProMED also runs something called EpiCore, which is aimed at getting around nations' reluctance or inability to report disease. Medical and veterinary workers who know some field epidemiology sign up for it, and then if ProMED hears a worrying rumor, it can ask them, privately, to check it out using a web platform that guarantees privacy. If you meet the criteria for membership, you might consider joining. Astonishingly, however, ProMED is supported by a few grants and voluntary donations, and can barely cover its costs. Could we maybe find it some independent funding? It is an embarrassment that such a mainstay of the global response to infectious disease has to regularly ask for donations. In April 2020, as Covid-19 raged, we all got emails signed by Marjorie Pollack pleading for readers to maybe pony up $25.

Finally, if we're going to take surveillance more seriously, there is one more area besides zoonotic hotspots we need to watch: labs. As we saw in the chapter on flu, research funding agencies in the United States and Europe are currently reluctant to permit experiments that, on purpose or not, make pathogens more dangerous, in efforts to find out how much threat they may pose, called gain-of-function research. The scientists argue, though, that we need to know whether some viruses can actually become more dangerous, and if so, what mutations to watch for. The US National Institute of Allergy and Infectious Diseases decided in 2019 to renew funding for studies of coronaviruses at the Wuhan Institute of Virology for just that reason.

The dilemma is always that doing the experiments without leak-proof containment might create the disaster we are trying to avoid. We can do research like that safely; we have been doing it for years. But I would argue that along with more funding for research, we must also fund better containment in labs and more, and more transparent, oversight to make sure researchers are working safely—and only on things that truly warrant the risk There have been too many known releases of dangerous microbes from labs already, and who knows how many that are less well documented.[36]

I've heard about too many cavalier experiments, or planned experiments, with viruses over the years, and read too many reports of accidental release, to believe scientists who insist nothing needs to change. There was a collective sharp intake of breath from the virology world in 2001 after a colleague on *New Scientist* broke the news that a lab in Australia had completely inadvertently created an extra-virulent strain of mouse-pox, a rodent infection related to smallpox, by giving the virus a gene for what they thought was an innocuous immune-modulating substance. Then, in 2003, at a meeting in Geneva, I heard an American scientist describe making and planning tests with an even deadlier version of that virus in a species of pox that humans can theoretically catch, although he was hoping the effect wouldn't manifest in people.

The late D. A. Henderson, who led the eradication of smallpox, was still alive then, and he was sitting next to me in the lecture hall, getting perceptibly angrier and angrier. Several other scientists in the room were also looking uneasy. When one asked what the researcher hoped to learn from the experiment that would warrant such a risk, a voice in the back said, "Nine-eleven." Apparently, we had to do this because terrorists might. I don't know if those experiments ever happened.[37]

One thing we can say about the virus that causes Covid-19, though, is that it was not made in a lab. In February 2020, as soon as they had had a chance to look carefully at the virus, Kristian Andersen of the Scripps Research Institute in La Jolla, California, and colleagues reported that, basically, virologists simply wouldn't have known enough to make a virus like this.[38] When a scientist makes a statement based on that kind of humility, I tend to believe it.

Proteins are strings of hundreds of a smaller kind of molecule called amino acids. The types and order of these amino acids determine the structure of the protein, which in turn determines what it can do—proteins are basically the tiny machines that do most of the processes of life. One spot on the Covid-19 virus's big external spike protein fits a spot on the ACE2 protein on human cells and binds to it so the virus can infect the cell. The binding site on the virus is a string of amino acids, Andersen and his colleagues admitted, that we wouldn't have predicted could bind to human ACE2. No virologist trying to build an artificial binding site would have chosen those. But it turns out they work just fine—and, far from having been designed to infect humans, could work better: subsequent mutations, says Rambaut, suggest the binding site is still adapting to us. The virus also has some apparently novel mutations initially seized on as evidence of engineering, but further analysis of wild coronaviruses, including new bat coronaviruses from Thailand, Cambodia, and Japan sequenced in March 2021, strongly suggests these arose naturally.[39] Most conclusively, as Ralph Baric later explained, SARS-2 has a basic structure unlike any coronaviruses previously isolated and studied in the lab. Creating a virus so novel, when we don't know what genes make a coronavirus good at spreading in humans, and without ever having seen a SARS-2 virus, would have been virtually impossible.[40]

Of course, a natural bat virus might escape from a lab studying such things. Conspiracy theorists make much of the fact that the bat virus with the closest known gene sequence to SARS-2, RaTG13, was collected in Yunnan, far from Wuhan, and insist

SARS-2 could only have got to Wuhan via the Wuhan Institute of Virology, which studies these viruses. But Yunnan, with its large, accessible bat colonies, is simply an easier place to collect samples from bats. In fact, the researchers initially found these viruses in bats near Wuhan in 2005, and there is no reason to think they are not still there. The same kinds of bats, probably with the same viruses, live across China and elsewhere in Southeast Asia.

Zhengli Shi provided a detailed rebuttal to then President Trump's accusations that Covid escaped from her lab, to the journal *Science* in July 2020. She pointed out that the sequence of SARS-2 doesn't match any live virus she's isolated and cultured—but then, there had been only three of those. The rest of the bat coronaviruses her lab has sequenced, including RaTG13, came from bat samples carried to the lab in closed tubes, from which the RNA was extracted—a process that kills the virus carrying it—under stringent containment.[41]

That containment is especially aimed at keeping lab workers from contracting any virus able to infect humans, then carrying and spreading it outside. Shi insisted none of her staff had antibodies showing any such infection. In January 2021, just as a WHO-led team arrived in China to investigate the origins of SARS-2, outgoing US Secretary of State Mike Pompeo stated that the United States had "reason to believe" researchers at the WIV had "symptoms consistent with both Covid-19 and common seasonal illnesses" in autumn 2019. He claimed these "revelations" undercut Shi's statement that there was no infection of staff with a SARS-related virus.[42]

It's worth mentioning this incident—which received press coverage—because it shows the sloppy thinking that gives rise to what the WHO called the "infodemic" around Covid. In fact, all this says is that people in the lab had cold and flu symptoms during cold and flu season—something so likely that it is easy to imagine that the only "reason to believe" Pompeo had was the fact that it was autumn. Sadly, too many such weak, tendentious arguments have become many people's received wisdom about Covid-19. You cannot do too much debunking.

Other routes between bat and human seem much more likely. There is far more virus in actual bats, which are out there flying around, than in labs. As we saw in Chapter 4, Shi's team found people living near the bat cave in Yunnan *had* been infected with SARS-like bat viruses. It is hard to see how virus in bat fecal samples killed in high containment could be more dangerous than the live virus in bat feces in nature, especially as many people put the latter in their eyes, as traditional medicine. It also seems wrongheaded to blame the lab that has been trying to warn us about these viruses for the fact that the warnings came true.

There is one other possible lab escape we should look at, however. Unlike Shi's lab on the outskirts of Wuhan, the Wuhan laboratory of the China CDC was, when Covid started, only three kilometers from the seafood market linked to many early cases of Covid—and since then it moved to an even closer location. According to a paper posted in 2020 by two Wuhan scientists—but later taken down after apparent pressure—the lab had a colony of live bats for virus studies, collected by a technician named Tian Junhua.[43] Tian told the *Wuhan Evening News* in 2017 that he had collected ten thousand bats for virus studies,[44] and is credited with supplying tissue from a bat in Hubei to labs elsewhere in China, which discovered a novel hantavirus in it in 2013.[45]

According to Dan Silver, a "rapporteur" for ProMED who follows health-related events in the Chinese media, in June 2019, very shortly before Covid emerged, the Wuhan CDC lab tendered for a hazardous waste-disposal firm to remove nearly two tons of medical waste from the lab that dated back twenty-five years, apparently a clearout before the lab moved three kilometers to new premises that December. It was a bad time to be asking: two fires at local waste-disposal facilities in May had left only one firm in Wuhan licensed to handle it. It took the contract in late July, presumably carrying it out in subsequent weeks. This kind of lab waste includes experimental animal carcasses, which are typically stored in freezers until they can be safely disposed of. The CDC lab's bats, killed at the time or put in the freezer after earlier dissection, and their droppings would have carried coronaviruses, which, Chinese scientists told the WHO in 2021, remain viable in frozen animal tissue.[46] Shi told *Science* that "based on daily academic exchanges" she ruled out a virus escaping from the CDC lab.[47] And the WHO-China investigating team ruled out the China CDC lab as a source in February 2021, as it conducted no work to isolate viruses. But the virus need not have been isolated or the focus of studies to escape from improperly handled bat material. Could pressure to dispose of the lab waste, and limited waste handling capacity, have led to the release of infected material? An infected waste-handler or even animals rummaging in garbage could have released a natural bat-borne virus in a crowded city center. As I write, this has not been investigated.

It remains beyond question, though, that the SARS-2 virus was not created in a lab. In March 2020, an unprecedented statement by twenty-seven of the biggest names in emerging disease ran in the

UK's top medical journal, *The Lancet*. "We stand together to strongly condemn conspiracy theories suggesting Covid-19 does not have a natural origin," it said. Scientists from many countries had studied it, "and they overwhelmingly conclude that this coronavirus originated in wildlife." Calling the efforts of China's scientific community to deal with the outbreak and share their results "remarkable," they concluded, "We want you, the science and health professionals of China, to know that we stand with you in your fight against this virus. Stand with our colleagues on the frontline!"[48]

That wouldn't be a bad rallying cry to carry forward into the post-Covid-19 world. There's no point doing any pandemic planning and surveillance without organizing a collaborative, global response to whatever we find out is happening. For all of us, now, "disease anywhere is disease everywhere" is no longer a slogan that sounds like it's for some telethon, but a lived reality.

But as diseases have emerged, there has been an absence of globally joined-up thinking. Zika took the Americas by surprise in 2015, even though it had already moved eastward into the Pacific from Asia, causing more severe disease than it ever had before, and even though chikungunya, another mosquito-borne virus from Africa, made exactly the same trip in 2013.

When virologists in Wuhan and North Carolina found bat coronaviruses that made mice sick and easily infected human cells in 2013, they practically shouted about the pandemic concerns this raised. Scientists also did that in 2004 about the family of swine flu viruses that indeed went pandemic in 2009. Both times, nothing much was done to head off the risk or prepare a response.

The One Health Platform is an organization of scientists that tries to bring together a wide range of researchers in human, animal, and environmental health and people from governments and international organizations to take a wider look at global health security. It is as good a place as any to start talking about how the world should build operational, day-to-day activities that really address the threat of pandemics—not just spotting novel pathogens, but doing something about them.

In a paper ahead of the World One Health Congress in 2020, the organizers said that in "peace time," between pandemics, we need the following: surveillance and diagnosis of disease syndromes in humans and animals; the identification of new pathogens; development of diagnostics and mechanisms for distributing them; research into how new infections cause disease; drugs; vaccines; and communication between scientists, governments, and the public.[49] That last one especially tends too often to be forgotten.

But that's a true scientist's list—mostly about finding out, not about taking action. Action, after all, is the job of governments, not scientists. So we might add to that list an authoritative, international capability for deciding when all that investigation has in fact revealed a potential threat that requires a response—drug and vaccine development, active monitoring—and making the response happen. Now that it is wartime, maybe governments will start paying attention—and paying for that kind of preparedness. It has to be someone's job to organize the response we need to the warnings we will turn up as we redouble our watch on emerging pathogens—assuming we do.

So, about those drugs and vaccines. First, know your enemy; then, choose your weapons.

## So What Do We Do About Disease?

As of March 2021, CEPI was organizing research and trials on twelve candidate Covid-19 vaccines, including the Moderna and AstraZeneca vaccines that were among the first to be released, adding more candidates as they emerged, and helping to organize the COVAX facility, with the WHO and GAVI, for ensuring Covid vaccines reached poorer countries. The WHO helped both to organize trials of existing antiviral drugs to see if they worked on the pandemic virus and to revive work on anti-coronavirus drugs abandoned after SARS. Covid taught us a lot about organizing in wartime. But we need to look forward, too, now that the first rush of R&D for Covid-19 is bearing fruit, or we will always be playing a deadly game of catch-up. "We cannot drop the ball on other diseases," Melanie Saville, CEPI's head of vaccine R&D, told me in early 2021. Despite the huge diversion of global money and expertise to Covid, in 2020 CEPI also organized work against Lassa, chikungunya, and MERS.

But scary as those WHO priority diseases are, they aren't the only potential enemy. Even viruses known to cause mere common colds can suddenly go rogue. Like pneumonia, a cold is a syndrome, not a specific germ: at least two hundred viruses cause them. In 2005, a novel adenovirus, Ad14, appeared on US military bases after they stopped vaccinating for adenoviruses, which are common causes of colds among recruits. This one caused severe pneumonia in 140 known people in the United States, many of them young and healthy, and there were probably many more cases that were not tested. Ten of the 140 died.[50] By 2008, most people were exposed and immune, and it became just another winter cold virus—but viruses don't always settle down and behave like that. And we don't know about most of them.

Another massive silver lining to the pandemic, however, was the demonstration of how fast and effective RNA vaccines can be to respond to such unexpected threats. Two were the first to be licensed and administered against Covid, partly because they were faster to design and make than more classic vaccines containing actual viruses or their parts.

For more than twenty years, researchers worked on vaccines made of the nucleic acid coding for a pathogen's protein. The person who got the vaccine translated that into protein, and their immune system would learn to recognize it, as with any vaccine. The first encouraging prototypes were DNA vaccines, but there were concerns that the DNA might, somehow, insert itself into our own genes. The answer was to switch to messenger RNA, which cannot do this.

But there was too little profit to be made from vaccines generally, and too much regulatory inertia about licensing a whole new class of them, for any human RNA vaccines to ever reach the market. Those barriers disappeared fast when Covid appeared, and RNA sequences from SARS-2 were being looked at as possible vaccines within hours of the first sequence being posted.

The speed with which the vaccines were developed, however, also depended on years of work in the wake of SARS and MERS, which taught vaccine developers that the virus's spike protein induced the most effective immunity, especially the form it takes before the virus invades cells and changes slightly. Because of that, the first vaccines for Covid were aimed at the right target.

RNA vaccines now present vast new possibilities for rapidly dealing with new pathogens: the ultimate "Disease X" platform. The global financing and distribution chains created for Covid vaccines will help, too. Moreover, unlike previous kinds of vaccines made of whole viruses or their proteins, you can make any sequence of RNA, from any pathogen, in the same manufacturing plant, meaning a global network of vaccine factories might be put on standby for any novel disease emergency.

But we will need the kind of background research we had with coronaviruses to be able to address any new threat as fast. And development in peacetime is unlikely to move anything like as fast as Covid vaccines did, without the unprecedented billions of dollars poured into dealing with an emergency that threatened rich countries as well as poor ones.

One goal, though, could be vaccines against a whole class of potentially pandemic viruses. For example, given all the bat coronaviruses still out there that can infect humans but may not be affected by immunity to Covid, we could use a universal coronavirus vaccine. In March 2021, CEPI said it would award research funding of $200 million for projects to develop a vaccine in the next eighteen months that would work against variants of SARS-2, some of which might evade the existing vaccines; and also to start designing a vaccine against all betacoronaviruses, the same family as SARS-2. Vaccines for other families, such as the adenoviruses, even the ubiquitous rhinoviruses, might also be good to look into in case one goes rogue. Nor should we forget other kinds of pathogens, such as the sometimes lethal food poisoning bacteria *E. coli* O157:H7. This could be the dawn of a golden age of vaccines.

Immunologists caution, though, that preventing all respiratory infections, even mild ones—the long-sought cure for the common cold—might be risky. Mild colds are thought to keep our immune systems primed to respond to other viruses that might be worse. Prior exposure to the other coronaviruses that cause mere colds, for example, might turn out to be why some people have less severe symptoms with Covid. Another area in need of far more research.

Meanwhile, the one vaccine that desperately needs funding is one CEPI doesn't do, but was a priority the GPMB highlighted last year: a universal flu vaccine. You have heard this already, but it cannot be said too often: a flu pandemic is inevitable. A pandemic flu carries novel surface proteins to which many or even all humans have little or no immunity. By definition, we cannot make a vaccine to a pandemic strain in advance, as flu vaccines are made of flu surface proteins and we have no idea what surface proteins the next one will have: there are endless variations, and immunity to one does not give you immunity to the others.

We could cook up a vaccine after the pandemic strain emerges when we know what it looks like—in fact, this is the current plan, as it is pretty much all we can do. But even with the added speed and potential worldwide manufacturing capability promised by mRNA vaccines, doing that cannot protect enough people fast enough. Even now the quantities of classic flu vaccine we can make aren't too constraining: the world can make 1.5 billion doses of winter flu vaccines per year, which means that, in theory, it can make 6.4 billion doses of a pandemic vaccine.[51] There are more people in the world, but vaccine experts have told me we are unlikely to reach everyone initially, even if we make more. Why the difference, though, between ordinary and pandemic vaccine-making capacity? In most seasonal flu vaccines, one dose contains 15 micrograms of the H protein from each of the three strains of flu circulating every winter: the H3N2 that emerged in the pandemic of 1968, the H1N1 that emerged in 2009, and the dominant strain of influenza B. Some vaccines have four strains, with an extra B. In theory, since a pandemic vaccine is targeted at just the one strain, a pandemic vaccine only needs 15 micrograms of that specific H. So when vaccine production lines switch from seasonal to pandemic vaccine, they are capable, in theory, of churning out enough flu virus for three to four times more individual shots than normal.

However, we might turn out to need far more of the pandemic virus's H protein to goad our immune systems into mounting an immune response, in which case we will be able to make far fewer doses of vaccine. That happened with some experimental vaccines for H5N1 bird flu. And for a novel virus people often need two doses each of a vaccine, given weeks apart, to be protected, as has been the case with Covid. Younger people who had never encountered the H1 on the 2009 pandemic flu, for example, needed two doses of that vaccine, while older people, who had seen it before, needed only one. So did people who had already had Covid when those vaccines emerged. But as we saw with Covid, two doses take time to administer, and also mean we need twice as many doses as we would for a one-shot vaccine.

Or, the outlook could be sunnier. We might be able to use an immune-stimulating chemical called an adjuvant to make small doses go further: several good ones have recently been developed, one of which is being used in candidate Covid-19 vaccines. Vaccine researchers have also made doses go further by using tiny micro-needles to inject the flu protein into our skin, instead of deep into a muscle as we do now. Skin is crawling with immune cells that can make the most of a tiny amount of vaccine.

At least we can now make four times more standard flu vaccine than we could in 2006. One reason is because, as concerns over bird flu mounted back then, poor countries worried about their access to vaccines in a pandemic. So the WHO launched a campaign to increase vaccine manufacturing capacity and put it in more poor countries.

But all of these plants use the standard process for making flu vaccine, growing flu virus in eggs, which takes six months to produce enough—and that's if the vaccine virus grows well. In the 2009 pandemic, there was no vaccine available until the end of the autumn wave, partly because the virus initially grew slowly. If that H1N1 had turned extra virulent in the autumn wave like its forebear in 1918, the fact that vaccines came too late would have been disastrous.

So despite all this effort to make more standard vaccine, we still probably can't produce enough, fast enough, to save a lot of people if a really lethal flu hits. There are a few proposals for growing made-to-measure flu vaccines faster—for example, by producing the required flu proteins in tobacco plants. RNA vaccines, as we mentioned, are fast to design and make and could be a game-changing way forward—if vaccine makers will develop and test an all-new mRNA flu vaccine that could render their existing vaccine plants obsolete, and as importantly, will build new plants to make large amounts of it at short notice, worldwide, preferably before another flu pandemic starts.

But the easiest solution, and the Holy Grail of flu, is a universal flu vaccine. Scientists have been working on this for some twenty years.

In theory, we could use bits of the flu virus that do not change, either year to year or between families of the virus, to immunize ourselves to all flu once and for all. Our immune systems mostly ignore these "constant" regions of the flu virus, seduced into making more antibodies against the big, obvious exterior of the H protein—which is one reason flu viruses have it. The hope is that if we were more strongly immunized against these constant regions, our immune systems would attack any flu virus we encounter.

We could give it as part of people's routine immunizations, to stop them getting ordinary winter flu and also any pandemic flu that might emerge, even though we don't know what that exact virus will be. We could stockpile it for people who hadn't already been vaccinated when the pandemic hit. Several candidate vaccines have passed safety tests and seem to induce the right immune reactions.

However, only one, made by the Israeli firm BiondVax and containing nine unvarying stretches of protein from flu, has so far managed to get funding for the expensive, large-scale trial needed to see if it works—and that trial was a disappointment.[52] It has been hard for companies developing these vaccines to find enough funding, for the usual reason: it would not be profitable for a company to make such a vaccine, as people would need only one shot or a few shots during the course of their lives. But no one other than the big companies can fund and organize such large-scale trials. The joke for years in flu circles has been that a universal flu vaccine is always five years away.

The GPMB called on governments to at least set a timeline, by September 2020, for developing a universal flu vaccine. It didn't happen. There has been a steady buzz of low-key research for years, and a universal flu vaccine does seem almost in reach—and mRNA vaccines open new possibilities here as well. How soon the world will buckle down on anything but Covid-19, though, is an open question.

It seems like all we need to finish the job is an intensive burst of coordinated research, large-scale trials of the best universal vaccine options, and money to build manufacturing plants for the winners. Covid-19 is showing we can do all of this when we need to. The start-up companies that do the initial vaccine research can't do it. If there was ever a case for public spending on a public good instead of leaving things to a market that simply cannot do this job, universal flu vaccine is it.

In fact, that, too, could be a silver lining to the dark cloud of Covid-19. Public-private partnerships have been starting to proliferate around needed but nonprofitable medical technologies like medicines and vaccines ever since big philanthropies like Gates—actually, ever since Gates—got involved in R&D for the diseases of poor countries in the 2000s. Now that might accelerate.

The press has been awash with commentaries claiming big government is back, as only governments can rescue the various industries that are going under during lockdowns and provide emergency income for people who have lost their jobs to social distancing. As a result, the idea of public investment for the public good seems to be returning from the wilderness to which neoliberal economic theories consigned it from the 1980s on. A lot may depend on how governments decide to handle their post-pandemic debt, but many voters may prefer their tax money go to accessible health care and pandemic preparation than to some other kinds of investment.

It is accepted that governments invest in public goods—roads, schools—to provide the infrastructure that allows private enterprise to flourish, at least in theory. Market failure means we have no universal flu vaccine or effective antiviral drugs, and we're losing our antibacterial drugs to resistant bacteria—we'll get to that in a moment. If a government really wants to support its industry, there's a case to be made for keeping its workers and consumers alive.

Of course, it's not only flu that should worry us. Another useful thing to have, if it is possible, would be a vaccine platform we can use for any virus that emerges: the WHO list's Disease X. One plan is to have an already tested, safe vaccine technology we can customize with a bit of the new virus, so it can be deployed with minimal further testing. We did that for Covid-19, partly because we had precedents. mRNA is the obvious ultimate anything-vaccine: you just have to change the genetic sequence to the pathogen you are targeting, in theory. But there are more conventional alternatives.

Two Ebola vaccines were developed amid the burst of anxiety and funding that followed the 2001 anthrax attacks in the United States, when it was feared Ebola might be used as a bioweapon. Work on them petered out as that burst of anxiety did, along with the funding. Also, with no sizable Ebola outbreak, the vaccines couldn't be tested.

Then Ebola struck West Africa in 2014, and big vaccine companies, to their credit, stepped up and organized trials. One of the vaccines, originally developed by the Public Health Agency of Canada, was 95 percent to 100 percent effective at stopping Ebola in contacts of cases. It is now called Ervebo, the first Ebola vaccine ever put on the market. Another vaccine was tested in the 2018 Ebola epidemic in the Democratic Republic of the Congo and was licensed in July 2020. In January 2021, the WHO, UNICEF, International Red Cross, and Doctors Without Borders started a stockpile of Ervebo for emergencies, but it will take several years before they have their target half-million doses.[53]

The point for future pandemics is that both vaccines consist of benign viruses (which, unlike Covid, carry their genes as DNA) and are fitted with an extra gene for an Ebola protein. The person vaccinated makes the protein, just as the person who gets an mRNA vaccine does. But these viruses are not as fragile as mRNA vaccines, which must be kept at very cold temperatures, because DNA is less fragile than RNA. In addition, the virus carrying it—the "vaccine platform"—gets the immune system's attention, possibly more effectively than every mRNA vaccine might. The Oxford-AstraZeneca Covid vaccine was of this type, based on the chimp adenovirus used in the Ebola vaccine approved in 2020.

The hope eventually is to have just such a tested vaccine platform ready so we can drop in a new protein from any surprise virus that emerges and have a vaccine quickly. CEPI had a pre-Covid goal of having one ready for tests in people within sixteen weeks of a new pathogen being detected. The AstraZeneca vaccine was in early human trials ten weeks after Zhang Yongzhen posted the first public SARS-2 gene sequence, thirteen weeks after China alerted the WHO to an outbreak in Wuhan.

There's a further complication, though. We can design and test as many vaccines as we like for emerging diseases and for pandemic flu, but before Covid it wasn't clear where we would make them in the large quantities required. RNA vaccines may be manufactured in a one-size-fits-all plant, but vaccines made of other kinds of whole viruses or proteins are more particular. There will never be enough demand for ordinary seasonal flu vaccine, industry insiders admit, to warrant enough manufacturing capacity for all the vaccine that would be needed in a flu pandemic, for example. How do we become equipped to make vaccines for a threat that has not yet materialized?

## So What Do We Do About Disease?

We solved the problem as best we could for Covid, because the world was facing an emergency, and even then vaccine was made more slowly than it might have been, partly because patents on vaccines and manufacturing techniques limited the companies that could make them: Doctors Without Borders, among others, called for scrapping them.[54] But the point of pandemic preparedness is getting these things ready before a pandemic happens. For example, we can brew up enough vaccine in labs to do safety tests in humans. But what if we have a Nipah vaccine we know is safe and want to use it in Bangladeshi villages where people are dying of Nipah to see if it saves lives? We might well need more than a lab can make.

We can't build a vaccine factory just for that vaccine if we don't yet know it works. There is little spare vaccine-making capacity. If the 2014 Ebola outbreak had needed more vaccine for trials than manufacturers managed to squeeze out with the little spare capacity they could drum up, production lines making important childhood vaccines would have had to be switched over. Fortunately, they didn't need it.

We might build vaccine factories just to have such spare capacity. But they aren't easy to keep in reserve, say experts at the WHO. You can't build one just for emergencies: it has to be working to keep its staff and processes up to scratch. In March 2020, the American Enterprise Institute called for a dedicated program to develop "flexible platforms" for producing drugs and vaccines for a new pathogen "in months not years," including "flexible manufacturing capacity to scale up production to a global level in an emergency."[55] Covid gave us some experience doing that. The question now is whether we can turn that into an ongoing, peacetime capability for pathogens that may not—yet—be an emergency.

We will also need to find ways to distribute the drugs and vaccines we create so all parts of the world have equitable access. We dealt with that for Covid in April 2020, when governments, the WHO, and philanthropies launched the Access to COVID-19 Tools Accelerator, a funding program that included the COVAX facility, which allowed rich countries to help finance Covid vaccines for poor countries and, in theory, guaranteed equitable access.[56] With luck, that could establish a precedent for other public-good drugs and vaccines in the future.

With less luck, we could see a repeat of the Trump administration's refusal to join COVAX. The injustice of this was evident, and his successor, Joe Biden, joined COVAX on his first day in office. Then in May 2021 he backed an international call to suspend intellectual property rules for the mRNA vaccines, so they could be manufactured by more vaccine producers in more countries and supplies would be available faster. This was partly because rich countries were undercutting COVAX by striking deals with vaccine companies to build their own stocks of Covid vaccine in addition to their contributions to COVAX, and because only companies with patent rights could make the vaccines, especially the most effective ones based on mRNA, this slowed production and massively delayed the arrival of vaccine in poor countries.[57] By June 2021, said the WHO, only ten countries had given three-quarters of the vaccine doses administered, and rich countries with 16 percent of the world's population had bought more than half the available Covid vaccine. This wasn't a transient effect as vaccination got underway: purchasing agreements had locked in that inequity until 2023.[58]

The hostile global competition for vaccine access was dubbed vaccinationalism and seems unlikely to disappear without a concerted international effort in peacetime to institutionalize the norm of sharing lifesaving technologies—possibly by permanently mandating changes to intellectual property rules that restrict manufacture of vital vaccines, tests, and other tools during a medical emergency. This will be especially needed if the next pandemic is something more lethal and the scramble for vaccines becomes all the more intense.

A failure to do that would be tragic. Equitable access is not only the ethically correct thing to do, but it is also simple self-interest, even for the richer countries that can get all the vaccine they themselves need.

In November 2020, the Rand Corporation, a prominent US think tank, calculated that even if all but the poorest countries were vaccinated, the world economy would still lose $153 billion a year, most of it from rich countries, to the continued drop in global trade resulting from the continued pandemic. If the rich countries, however, paid the full cost of vaccinating the poorest ones, it would cost them only a fifth as much as they would otherwise have lost, an amazing bargain.[59]

And that could be a conservative estimate. In a more wide-ranging assessment, the National Bureau of Economic Research in the United States, a leading economic think tank, calculated in January 2021 that because countries' trade and economies are globally interdependent, then if vaccination continued its slow pace in poorer countries, even if rich countries vaccinate themselves they'll still lose $4.5 trillion in 2021, especially from sectors like construction, textiles, and retail. But if rich countries gave the ACT Accelerator the $272.2 billion it needs, they would prevent losses from their own economies worth 166 times that—making global vaccines not just a bargain, but in the words of the International Chamber of Commerce, "a major investment opportunity."[60]

And that's just dollars; as always, things get much more dire when you take immunology into account. Let's say one part of the world vaccinates its people, but another part cannot, and the disease continues to spread there—even for only, say, a year, the minimum time difference that seemed likely, in March 2021, between getting rich and poor countries vaccinated.[61] The virus in the diseased part of the world will continue to evolve, as indeed SARS-2 evolved just during 2020. Soon it might well become one the vaccines don't work against.

We have all seen how hard it would be to keep that virus out of the vaccinated area—and when it arrives, the vaccinated population is back to square one. We truly are all in this together. As Rand puts it, "Enforceable frameworks are needed for vaccine development and distribution. Countries need to be bound by an agreement."[62] So far that is only partly happening, even for the present emergency of Covid. The WHO's independent investigative panel lamented in January 2021, "We cannot allow a principle to be established that it is acceptable for high-income countries to be able to vaccinate 100% of their populations while poorer countries must make do with only 20% coverage."[63] That 20 percent may be overly diplomatic: in late January 2021, as vaccination took off in rich countries, only twenty-five people were vaccinated in all Africa, all of them in Guinea.[64]

There is a more speculative area of therapeutics we should consider. Covid-19, like SARS, seems to kill partly by triggering excessive inflammation, which can kill even well after your immune system has rid you of the virus. Normally, inflammation is a general activation of the immune system that gets rid of infections—in fact, SARS-2 defends itself by inhibiting one of the body's early inflammatory signals—but it can get out of balance. The reason older people and people with underlying conditions such as diabetes, high blood pressure, and even obesity have a higher rate of severe disease and death with Covid-19—and flu—is because all those conditions, including aging, involve chronic inflammation. The virus triggers more inflammatory reactions on top of that, and things spiral out of control.

Inflammation is fiendishly complicated, so it's difficult to tinker with, but some drug developers are now looking at ways to tackle excessive inflammatory responses themselves as a way to limit both the chronic underlying conditions and the impacts of infectious diseases, and maybe even some aspects of aging. Research in the UK showed in 2020 that a classic anti-inflammatory drug, dexamethasone, helped patients late in the disease when the virus is gone but inflammation can rage out of control. In January 2021, two more anti-inflammation drugs, originally developed for arthritis, seemed to cut Covid deaths in intensive care by a quarter.[65] A space to watch.

GPMB also called for work on broad-spectrum antivirals, analogous to the broad-spectrum antibiotics that kill a wide variety of bacteria. Such drugs can, in theory, be used to knock down any unexpected virus that emerges. But wide-spectrum antibiotics provide a cautionary tale: because they kill many kinds of bacteria, they also promote widespread antibiotic resistance. Viruses can develop resistance, too. We have two families of antiviral drugs for flu, and genes for resistance have emerged for both of them. One kind is in the troublesome H5N1 bird flu.

Thankfully, the antiviral Tamiflu still works against most flu, and it is stockpiled in some countries in case of a flu pandemic. But this illustrates another kind of threat. There has been a denialist crusade against the drug and the pandemic stockpiles, based on claims that the manufacturer's drug trials show it doesn't do much against ordinary winter flu—claims that are themselves scientifically questionable.[66] One critic told a British parliamentary committee there was no evidence that Tamiflu was better than "a stiff whisky."[67]

In fact, there is plenty of evidence. The drug is stockpiled for pandemics, not ordinary winter flu, and it is used in a pandemic to stop people from dying of severe viral pneumonia, similar to what Covid-19 causes, which happens in flu pandemics far more often than in normal winter flu. The drug trials were aimed at determining whether the drug affected ordinary flu, not severe pandemic flu, as you can't test a drug against a disease that isn't happening at the time. But Jonathan Van-Tam of the University of Nottingham—who became a popular public voice for pandemic science as deputy chief medical officer for England in 2020—found that among 168,000 people with flu severe enough to need hospitalization in the 2009 pandemic, people who received Tamiflu within two days of falling ill were half as likely to die—a significant effect.[68] We would have loved to have had a drug like that for Covid-19 when the pandemic started. Yet the crusade against the flu drug continues, with a lawsuit launched against the manufacturer, Roche, in the United States in January 2020, for allegedly "bilking" the US government of the money it paid for its stockpile, still dragging on in 2021.

## So What Do We Do About Disease?

This seems like a good time to address the elephant in the room whenever we talk about any aspect of our future health, including the risk of pandemic viruses: antibiotics, the drugs that kill bacterial infections. No one really expects a bacterial disease to stage a pandemic, although I am increasingly reluctant to rule out anything involving the living world. But bacteria aren't as rapidly evolving or communicable as many viruses—there's a reason we call something speeding across the internet "viral."

Yet antibiotics play a role in a viral pandemic like Covid-19. Any use of antibiotics promotes bacteria that can resist the drugs, so for many years the WHO has pleaded with countries to use them only for infections where they are needed and will actually work. But early in the pandemic doctors treated most patients hospitalized with Covid with antibiotics to prevent any coinfection by bacteria, especially patients on ventilators, which increase the risk of bacterial infection. The precaution was reasonable when Covid started, but by late 2020, evidence was emerging that only 15 percent of patients really needed such antibiotics, yet 75 percent were getting them. The WHO issued warnings, but in 2021 the overuse was continuing.[69]

Making matters worse, the WHO European office found many people were taking antibiotics in the misinformed hope they would ward off the virus.[70] It didn't help that some people falsely promoted antibiotics as cures—for example, Donald Trump's advocacy of azithromycin, which has no effect on Covid.

One reason for the confusion is that antibiotics would assuredly be needed in a flu pandemic. A third to half of the millions who died in the flu pandemic of 1918 are thought to have been killed not by direct viral pneumonia, but by the bacterial pneumonia that often follows flu. Historians often reassure readers that 1918 could never happen again in our modern world because now we have antibiotics.

This easy assumption that we will always have effective antibiotics in such a case always makes me shudder. A growing number of such infections already resist antibiotics, and we get more resistant bacteria the more antibiotics we use. The massive use and misuse of antibiotics to desperately treat bacterial complications in a flu pandemic could accelerate the process. It isn't yet clear how bad Covid has been for this, but it is probably happening.

Many antibiotics come from microscopic fungi in soil, which use them in their constant war with soil bacteria. Accordingly, the bacteria have developed genes for proteins that block or demolish the fungal antibiotics. And bacteria swap genes like foodies share sourdough recipes. Probably more.

If you expose bacteria to an antibiotic, some may well have a gene for resisting it, or even several—bacterial genes travel from cell to cell in packs. As we have used more and more antibiotics, the bacteria that survived were the ones that had these genes and could defend themselves, so they have become more and more common—there are few clearer illustrations of how evolution works. Antibiotics are prescribed at doses and over times that should kill all the bacteria, but even then, resistance can emerge. Infections resisted the very first medical antibiotic, penicillin, deservedly called a wonder drug, just three years after it was first widely used in people.

But misused antibiotics, as when patients demand antibiotics for ordinary flu or use it to ward off Covid, promote resistance even faster, because you aren't using the drug to kill off a target population of bacteria, but bacteria are still being exposed randomly, and the resistant ones survive. The low doses of antibiotics given to cattle, pigs, and poultry to make the animals grow faster do this, too. There have been choruses of denial from livestock industries, but the science is clear: this practice contributes to antibiotic resistance in bacteria that cause human infections. Researchers have tracked it, as they say, from farm to fork. The European Union has banned antibiotic growth promoters, which demonstrates that modern animal production doesn't need them, but the United States has been slow to give them up, and the drugs are used massively as livestock production booms in South America, Asia, and Africa.

We really don't want to lose antibiotics, especially if we expect more pandemics, especially of flu. Few people realize what a huge difference they have made to human welfare. Indeed, few people reading this have not at some point—probably several points— had their lives saved by them. Have you ever had an operation, even something ordinary like having your knee fixed or appendix removed? You needed antibiotics to stop the bacteria that invaded your opened-up body. Have you or a loved one ever had cancer treatment? Cancer drugs suppress immunity, so you need antibiotics, or the bacteria that grow as your immunity falls can kill you.

Have you ever had an abscess, an injury, dental surgery, a bout of bacterial pneumonia, a sexually transmitted disease like gonorrhea, or a common urinary tract infection? Antibiotic resistance affects all of them, and there are now cases of those last two that resist all known antibiotics—they're incurable. Mothers and babies used to die in droves from bacterial infections around childbirth—and where they can't get modern medical care and effective antibiotics, they still do. That used to be normal for all humans: a tiny cut could result in gangrene or sepsis. Now, if you get an antibiotic-resistant infection in a cut, that can happen again.

So invent better antibiotics, you say. Exactly. But for the same reasons we don't have better flu vaccines or treatments for coronaviruses, we don't have many new antibiotics hitting the market. In an investigation I did for *New Scientist* in 2019, I found that unlike a few years previously, researchers and research funding agencies had taken up the fight, and a lot of new kinds of antibacterial drugs are in development, including clever new approaches like harnessing viruses that infect bacteria.[71]

But industry experts warned me that however good these were, they were unlikely to get the $1 billion worth of trials in people a drug needs to be marketed safely. Antibiotics, like flu vaccines, are not terribly profitable. People take them for only a week, whereas they go on buying blood pressure pills, arthritis drugs, or Viagra for years.

Moreover, new antibiotics should not be widely or aggressively sold but ideally saved for infections that resist existing drugs to avoid encouraging resistance to the new drug. Yet it is when a drug has just come on the market that companies most need to start selling it heavily to recoup their R&D investment. And even when new antibiotics are the best treatment and should be used, doctors tend to try older, cheaper ones first. As mentioned earlier, there are proposals for methods of repaying companies through mechanisms other than sales, but none has been widely applied yet.

## So What Do We Do About Disease?

Thus, many Big Pharma companies have abandoned R&D for antibiotics: eighteen developed them in 1980, now only six have any kind of program, and they may not last. Fortunately, small start-up companies are developing new antibiotics, but in 2019, several went bankrupt. One in California, Achaogen, had spent the requisite $1 billion getting a drug, Zemdri, onto the market that cures antibiotic-resistant urinary tract infections. However, the company needed more money to market the drug and do more studies. Investors saw little chance of profit and declined. The drug was bought by another firm, but the company is gone, the researchers are off doing other things, and fewer inventors with a great idea for a new antibiotic like that are now likely to pursue it.

So we risk losing antibiotics. In 2014, a blue-ribbon commission in the UK reported that seven hundred thousand people a year were already dying worldwide of antibiotic-resistant infections, fifty thousand just in Europe and the United States—but by 2050, this number could jump to ten million a year, more than the number who die of cancer today and more than seven times the number who die in road accidents every year. That many deaths would be horrendous, besides slashing trillions from global GDP.[72] A pandemic virus would be proud of those numbers, if proud was something a virus could be. If we agree we need to care about pandemic risk, then we need to care about this.

This is partly because it really matters to our risk of viral pandemics. The 2014 report forecast economic damage from more and more hobbled health care resulting in severe domino effects as we increasingly lose antibiotics: less money overall means less for new treatments for emerging disease, for stockpiles in case of a pandemic, for disease surveillance.

The good news is that the solutions we need to encourage new antibacterial treatments are largely the same as those we need to encourage drugs, vaccines, and tests for potentially pandemic viruses. You guessed it: what they all have in common is that, for all its virtues, the free market on its own cannot make them happen.

That means public investment. An organization called CARB-X has $500 million to invest between 2016 and 2021 to accelerate development of new antibiotics, similar to what CEPI does for vaccines. "One lesson from COVID-19: we need to invest today for tomorrow's pandemic," says Kevin Outterson of CARB-X. "What would a COVID-19 treatment or vaccine have been worth in 2018? Zero—the market would have seen no value at all. What would they be worth today? Off the charts. That is true for pandemic viruses and also for drug-resistant bacterial infections."

But like CEPI, CARB-X only gets products as far as initial trials. The problem is the big expensive push they then need to get onto the market. With Covid-19 drugs and vaccines, the need is obvious, so the money has been found. With antibiotics, the need is clear, too, but even so, few new ones are in large-scale clinical trials. We may need to get creative.

In July 2020 the pharmaceutical industry put together a $1 billion fund, the AMR (antimicrobial resistance) Initiative (www. amractionfund.com) to put promising drugs that have got through initial trials onto the market. Another model is the US Project Bioshield, set up in 2004, after the anthrax attacks, to help develop vaccines and treatments for germs that might be made into bio-weapons. This was established not to provide initial R&D, but to get companies with a promising product through the "valley of death"—the long wait during safety and effectiveness trials before they can sell.

Another US agency, the Biomedical Advanced Research and Development Authority, or BARDA, took it over in 2006, expanded the remit to pandemics, and has invested $1 billion in antibiotics. Unfortunately, that included $124 million for Achaogen, which didn't keep the firm from going under. But that just shows we need more consistent follow-through. Amesh Adalja observes that Bioshield cost in the low billions of dollars, far less than Covid-19 will cost. Preparation is always cheaper than reparation.

One can imagine some mechanism like that set up internationally for pandemic threats, like CEPI, with enough funding to get drugs and vaccines for potential pandemics as well as new antibiotics through all the requisite trials. Guarantees from governments to buy the successful results would help, and working globally—which, after all, is what pandemic diseases and resistant bacteria do—would bring economies of scale. It seems at least worth a try. Antibiotic resistance is effectively already a pandemic. If we don't include antibiotics in any grand plans for pandemic preparedness, we are setting ourselves up for failure.

There will be lots of strife long before the dust settles on Covid-19 over what worked and what didn't and how this or that solution was too late or ignored. We should turn that to constructive use, if we can, to hammer out plans that actually will work next time. We have already learned a few things.

We need better rules for using the high-tech methods and smartphone apps now being devised to trace contacts and enforce lockdowns, when they are needed. Contact tracing especially is difficult and expensive but can be vital, as we have learned, and it can be made vastly more effective using apps like the ones now being developed—these plus rapid tests can in fact be good ways to ensure painful lockdowns are not needed by allowing social distancing only of the infected and exposed. There must be ways to use these technologies without demolishing people's privacy or enabling authoritarian streaks (or worse) in some governments. Watchdogs with serious teeth will be essential.

The lack of medical equipment has rightfully received much attention. It is to be hoped that countries will stockpile the things we desperately needed during this pandemic: ventilators, PPE, and masks for the public. Experience is a good teacher: the Canadian province of Ontario, where Toronto was hit hard by SARS in 2003, had extra ventilators stockpiled that it used for Covid-19. I wonder how often Ontario had to defend that stash against critics calling it a needless expense now that SARS was gone? In 2006, California governor Arnold Schwarzenegger made similar stocks of pandemic supplies; they were lost to a budget crunch under another governor in 2011, and the state ran short of ventilators when Covid-19 hit.[73] Have we learned our lesson?

Adequate testing has also been a major, unexpected issue. One remedy could be pre-pandemic contracts with test manufacturers, to be activated in an emergency. South Korea signed agreements with test makers on the fly early in the Covid-19 pandemic, which let the companies develop and start mass-producing tests in days and also to use them and put them through validation trials at the same time. That famously enabled South Korea to contain the virus fast. Other countries could follow their lead and prearrange such agreements with test companies and other suppliers of emergency pandemic goods before the next one hits.

Meanwhile, individuals should not have to choose between spreading a pandemic and feeding their families. Even before this pandemic, research found that paid sick leave, so employees do not engage in "presenteeism"—going to work at any cost, including when sick—ultimately saves companies money. In the Covid-19 pandemic, it saved lives. Guaranteeing that right for working people even in the gig economy is entirely possible, says the UN's International Labour Organization, and would increase resilience to infectious disease in places where sick leave is not already an unquestioned labor standard.[74]

Of course, all these ideas—pandemic stockpiles, global surveillance, flu vaccine, sick leave—cost money. But the big thing we can do now as a result of Covid is put that cost in context. Those who warned about pandemics before Covid struck were already trying. The GPMB assembled some sobering figures. Zika cost the Americas some $20 billion, including care for the many handicapped children it left. SARS cost the world $40 billion. The 2009 flu pandemic cost $55 billion.

The 2014 Ebola epidemic in Africa cost the world $53 billion.[75] A lot of the financial loss to the three countries directly affected by Ebola was due to medical services not performed—vaccination, childbirth, malaria treatment—because of the disruption of medical services. That has also happened with the Covid-19 pandemic, as parents avoided getting children vaccinated, endangering polio eradication[76] and leading to severe measles outbreaks.[77] The disruption in medical service and people avoiding hospitals led to a "catastrophic" fall in cancer care across Europe,[78] reduced HIV care in the United States,[79] and disrupted maternity care worldwide,[80] contributing to more than a quarter of a million deaths of children and mothers just in South Asia.[81] Cardiac arrests doubled in Paris and tripled in New York at the height of Covid-19 in 2020, partly because the disease itself causes heart damage, but also because people already at risk were afraid to go to hospitals.[82]

There are other kinds of cost. Epidemiologists at Imperial College London calculated in May 2020 that because medical services worldwide were preoccupied with Covid-19, deaths from HIV, TB, and malaria, which had been steadily declining, would rise as much as 36 percent over the next five years in low- and middle-income countries.[83] In December 2020, the Global Fund to Fight AIDS, Tuberculosis and Malaria calculated that there would be over half a million more HIV-related deaths in 2020 than in 2018, numbers last seen in 2008, demolishing more than a decade of progress. Malaria deaths rose by nearly four hundred thousand, putting the fight back twenty years.[84] In March 2021 the Stop TB Partnership reported that Covid's impact on TB—the world's deadliest, endemic infectious disease, killing some 1.5 million people every year—had eliminated twelve years of progress in diagnosis and treatment.[85]

And we are learning that disease is more expensive than we realized, as the social distancing and lockdowns required to slow Covid-19 have destroyed livelihoods, in some cases past recovery. In October 2020, the International Monetary Fund calculated that economic growth would fall in 2020, rebound somewhat as the crisis subsided in 2021, but then gradually slow due to deep, structural economic damage worldwide. "We have a projection of a permanent loss in output pretty much everywhere in the world," said chief economist Gita Gopinath. The cumulative loss in economic output compared with what had been expected before the pandemic was an eye-watering $11 trillion in 2020–2021, rising to $28 trillion by the end of 2025—maybe only $20 trillion if vaccines are rapidly made available.

That is "a serious setback to improvements in average living standards" everywhere, said Gopinath, in rich countries as well as poor.[86] We can—for now—speculate only about the effect this might have on political stability. But as a result of Covid-19, no one can dispute that the cost of pandemic prevention and preparedness is just a drop in the bucket compared with the cost of an actual pandemic. It is also clear that as pandemics are global, preparedness needs to be, too, and countries must coordinate and share efforts.

And yet, as of early mid-2021, it isn't happening. The EU is the world's most integrated bloc of countries, and it is having trouble coordinating its members' response to Covid. "I am not yet too optimistic about a joint EU preparedness strategy," says Ab Osterhaus. "We need more EU strategy and funding for better future pandemic preparedness—universal strategies, vaccines, antivirals, diagnostics, etc." Yet even government attention to pandemic flu preparedness, the one kind most had done something about, "has largely waned," he says. Seth Berkley at GAVI is one of those people you meet in Geneva who is very clear-eyed about the world after years of working internationally. In 2017, when I was writing about preparing for pandemics for *New Scientist*, I asked him how to make countries take the problem seriously.[87]

Seth is American, and he observed that the United States maintains three kinds of nuclear weapons—airborne, land-based, and submarine-launched missiles—so it will have one deterrent left if the other two fail. "The chances of that happening are very small, but we spend tens of billions of dollars a year to keep that running," he said. His point was that if it's worth investing $49 billion a year to ensure you can respond to an improbably successful nuclear attack, surely we could invest in improving our response to an increasingly probable pandemic.[88] Yet the whole world spends only $2.4 billion a year, 5 percent of the annual cost of the US nuclear deterrent, on the WHO budget.[89]

He also said something I don't think I really understood until Covid-19. "The really big problem is appreciating what is at stake," he told me. "If people understood the risk, they would want to be sure systems are in place to deal with it. The costs of doing that are trivial compared to the cost of ignoring it."

We all know that now. The only questions are whether we will forget about it again, after Covid-19, or whether we will finally be able, together, to act on it. Just to make it harder to forget, let's look at some of the scarier risks.

# CHAPTER 7

# Things Fall Apart

> So, it turns out the most important jobs
> are not the bankers, the brokers, or the
> hedge fund managers. It's the doctors,
> the nurses, the hospital porters, the A&E
> administrators, the bin men, the teach-
> ers, the carers, the supermarket shelf
> stackers. . . . Who knew?
> —T-shirt from 2020[1]

A few years back, I started looking at what would happen if another plague like the Great Pestilence of 1347 hit the world. Dubbed the Black Death by later scholars, the medieval plague killed a third or more of all people in Europe and spread into Asia. I was writing about potentially pandemic pathogens with high death rates and wondering, what if?

Now let me say right at the start that I am not even remotely suggesting that Covid-19 is showing any sign whatever of becoming that bad. But there are viruses out there that are a lot worse than Covid-19, and as we will discuss, it is not at all a foregone conclusion that they will start playing nice if they go pandemic.

In any case, what I discovered by asking what might happen if another Black Death hit us turns out to be very relevant to what we are already going through with Covid-19, even though it has been much less lethal. The link, as with many apparently intractable problems, is complexity.[2]

European civilization did not collapse from the mass death that started in 1347, even though the disease circulated, with smaller outbreaks, for the next three hundred to four hundred years. Some historians even think the resulting labor shortages shook up the rigid feudal system and triggered changes that led to the modern era. The key was the simplicity of the agrarian society it struck. In such systems, nine in ten people are subsistence farmers, producing just enough to feed themselves, plus a meager surplus that, in medieval Europe, fed a few aristocrats, churchmen, and towns. Most deaths took out a peasant, and therefore a producer, but also a consumer, so had little net impact on society, especially as consumers who produced no food, and depended on peasants, were taken out at about the same rate. Nearly everyone was a nonspecialist and could be replaced. Even kings were replaceable.

But as Joseph Tainter, author of *The Collapse of Complex Societies*, told me, around 170 CE the Roman Empire was hit by an almost equally deadly plague that he believes tipped Roman civilization into a death spiral. The Antonine Plague was probably smallpox, formerly unknown in Europe, and its initial impacts in cities resembled those of lockdown, as city dwellers trying to avoid infection abandoned markets and hoarded food, and businesses shut. The real difference from Europe in 1347 was the empire's large urban populations, not equaled until modern times, and the complex networks of grain shipments and taxes and armies that supported them. Losing a third of the population meant grain production and taxes fell, the army suffered, invaders Rome would once have easily repulsed made inroads, grain and taxes fell further, and so on. Ultimately cities—the *civis* that was Roman "civilization"—largely disappeared. Decline led to fall.[3]

The difference was complexity. Defined very simply, a complex system is one in which many elements interact closely and feed back on each other—change one bit, it changes another, which changes a third in a way that reverses the first change a bit but also affects a fourth, depending on what a fifth is doing. The important thing to know about complex systems is that they behave very differently from the linear, mechanical systems we are more familiar with, where if you put something in one end, you get a predictable response out the other.[4]

In a complex system, if you change one bit, you might get a completely disproportionate response you were not predicting, because you don't know the states of all the components at that precise moment or how they all affect each other. The famous butterfly effect, where the flap of a butterfly's wings in Brazil could set off a tornado in Texas, reflects early efforts to model the weather, a complex system where tiny differences in starting conditions can create huge differences in outcome.[5] These are called nonlinear effects. This happens in all complex systems. A large change can also have small effects—up to a point.

This matters because complex systems have a few more

universal properties. Complexity can only be maintained with energy. The laws of thermodynamics, the most fundamental laws of nature, make it quite clear that, in strict scientific terms, there is no free lunch. To maintain a system more complex than random atoms—you, for example—you need to spend energy on it. For you, this comes in the form of food. You process the energy and materials bound up in the orderly structure of your sandwich to build and maintain the orderly structure of you, except for some energy lost to friction in the system. And other living things, food or food workers, had to do the same thing to assemble the orderly materials in your sandwich. No free lunch.

Moreover, complex systems tend to remain stable in the face of the normal range of conditions they evolved to handle, a property called resilience: disturb the system, and complex, adaptive readjustments tend to keep it on an even keel. This isn't magic. Complex systems evolve over time by trial and error and tend to self-assemble rather than being designed from outside: the enormously complex network that manages to deliver an amazing array of food to a big city every day is an example. Resilience evolves in such systems for the same reason anything evolves: because it can, and it works.

But if you push such a system outside the conditions it evolved with, resilience can vanish. A small change can flip it into an alternative stable state—the famous tipping point. A straw breaks an overloaded camel's back. A few tiny bacteria that produce deadly botulinum toxin, hiding in that sandwich you just ate, can kill you. So can a virus.

Society is a system that gets steadily more complex, says Tainter, because whatever we do, we encounter problems we have to solve.

We start growing crops to have more reliable food, but the rains sometimes fail, so we dig irrigation canals. Those silt up, so we invent dredging. They silt up more, so we have permanent dredging crews, and they don't farm so we give them food other people grew. Disputes arise, so we invent ways to record who gave and got what, then a class of people to keep track, another to keep order, and they all must also be fed. You can see where this is headed.

More complexity costs more energy. Human history is a long saga of people learning to harness ever-increasing amounts of energy to maintain ever more complex, ordered systems, punctuated by periodic collapses—the Romans, the Maya, virtually every previous civilization—when they became more complex than they could maintain, with the energy and technologies they had, in the face of changing conditions. At that point, small stresses sent overstretched social systems into a rapid downward spiral, which ended with major losses of people and social organization, as one stable complex system made a rapid nonlinear descent to a less complex one. But after a setback, humanity always innovated and rebuilt, a little bigger and more complex than before.

This process is integral to how we should understand pandemics. We now live in the most complex civilization the world has ever seen and the first to encompass the entire planet. Many believe this makes us resilient to shocks. But, say the complexity theorists, the more complex systems get—the more tightly coupled their component parts, the faster and denser the communication and transport links that keep them all coordinated, the more closely each part relies on many other parts—then the more rigid the system gets overall, the less resilient, the more likely to collapse.

Moreover, complex systems—natural ecosystems as much as human societies—tend to become more efficient with more specialized components and fewer redundant linkages, because that saves energy—or its surrogate, money. Thomas Homer-Dixon, a Canadian expert in complex systems and author of *The Upside of Down*, notes that a mature forest may have one kind of bacteria fixing its soil nitrogen, whereas at an earlier stage of development, it had a dozen.

Similarly, protective medical gear and the active ingredients for common, emergency drugs used to be produced widely. Michael Osterholm is an epidemiologist who has studied the possible impacts of pandemics. He told me that now a few factories in China make nearly all of these vital supplies, as the global industry takes advantage of low labor costs and economies of scale. This is efficient. Hospitals rely on constant, just-in-time deliveries of these items, too: keeping stocks costs money, so this is also efficient. During the early days of the Covid-19 pandemic when much of China was affected, there were fears deliveries would stop, either because China needed more of these things than usual or because factories or shipping might shut down as employees were quarantined. If things had gotten much worse or the shutdown had lasted longer, they might well have. There were no alternate sources. Efficient coupling between parts of the system would have led to a breakdown.

Homer-Dixon says increasing complexity makes societies more resilient only up to a point. Connections between villages might mean one comes to the other's aid in an attack. But as the villages become more tightly coupled, both may suffer when one is attacked. A loose network absorbs shock; a tightly coupled one transmits it.

That has happened in the Covid-19 pandemic. Countries go into lockdown; people stop shopping, traveling, and producing; and the effects ricochet through a tightly coupled global economy. The global supply chains of money, materials, people, energy, and component parts that underpin industries falter and break. Airlines go under as they are not set up to weather even a temporary disappearance of travelers. Malaria worsens in Africa as insecticide and antimalarial bed net deliveries falter. Microcredit that underpins small businesses throughout the developing world defaults because payment collectors are locked down, causing ramifications throughout an economy.

The number of people facing starvation threatened to double in April 2020, warned the World Food Programme, even though the same amount of food was available. Lockdowns meant poor people, from tuk-tuk drivers to cleaners to street vendors, could no longer earn money to buy it—and this happened just as restricted global transport made it hard to get donated food to them.[6] In March 2021, UN food agencies said the economic disruption of Covid-19, combined with climate extremes and conflict—and locusts—had sent acute hunger soaring in more than twenty countries, with millions facing starvation.[7]

Just the fact that an outbreak in China went pandemic as quickly as it did is a testament to the tight coupling in our global system. For human viruses, the vector is people and airplanes. Scientists tracked this pandemic using computer models and databases of global air passengers. Alessandro Vespignani of Northeastern

University calculated that the countries at highest risk of importing a case of Covid-19 were in Asia, followed by Europe and North America; that is exactly how the virus traveled. Vittoria Colizza of Sorbonne University in Paris used the passenger data to calculate that the African country most likely to import a case was Egypt, followed by Algeria.[8] Those countries, in that order, got Africa's first cases.

The fact that the world is a complex system helps explain how this pandemic happened. Basically, it means our system has a management problem. People tend to see things in a simple, linear way. That's not a criticism—we can't usually control anything but a few, simple, direct interactions within our social, political, or economic spheres. So faced with a problem, those are the solutions on offer. We cannot always anticipate how that will work out with the rest of the complex system. Chinese medical authorities said: we had a close call with SARS and we have bird flu, let's have doctors signal any unexplained pneumonia, and we'll see any clusters of cases faster. Scientists said: we have a problem with animal viruses infecting people, let's swab a lot of animals and see what they're carrying. Pandemic planners said: if we have a flu pandemic, we'll need ventilators and masks, so we'll stockpile them. All great ideas, and it was a good thing people did or tried to do all of them.

But complex interactions took an unexpected hand. In December 2019, when it was clear the unexplained pneumonia wasn't bird flu, for some reason Wuhan doctors were told not to use the alert system. Scientists found a virus much like Covid-19 in bats and warned of its pandemic potential, but that didn't nudge research agencies to fund renewed work on coronavirus vaccines. The 2008 market crash—itself a textbook product of complexity and tight coupling in the global financial system—triggered government cuts that tightened health budgets. Then the 2009 flu pandemic was less than catastrophic. Result: hospitals, with few exceptions, did not have the pandemic stockpiles envisioned in 2006.

Western countries have been talking about pandemic preparedness since bird flu rang alarms in the early 2000s. This was especially true in the United States, which as we have seen was widely—and wrongly—expected to be the country best prepared for something like this. But when Covid-19 hit the United States, the plan was largely abandoned, while unexpected complications set in everywhere. Health care workers didn't have enough protective gear and ended up sick or in quarantine, further restricting already limited health care. Insurance rules meant people initially couldn't afford to get tested. For weeks, they couldn't get tests in any case because of problems with the one test permitted from the CDC in Atlanta. Employees with no paid sick leave came in to work, hoping it was just flu. The virus spread earlier and farther than surveillance systems could detect, partly due to years of cuts to public health, partly because there was far more symptomless spread than we initially suspected. In some societies, including the United States, many people actively rejected lifesaving fixes from masks to lockdown to vaccines, claiming it violated personal freedom, even though the personal freedom to spread lethal germs to other people is not permitted in anyone's laws.

Pandemic planners didn't see all of those coming, although they warned about many of them. But no one could change enough of the system to head them off, and when a severe pandemic didn't materialize after years of warnings, leaders lost interest. We can't prepare a complex system for events like pandemics with small, linear solutions to local parts of the problem; we also can't prepare when we lose focus on any given risk after a few years.

So is there any hope? Actually, there is. We will look at potential solutions in the next chapter. For the moment, though, let's see how high the stakes really are. If we know how bad things can get, we'll know just how much we need to get cracking on those solutions—preferably before something worse goes pandemic.

For starters, how bad can a pandemic pathogen get?

That isn't as simple a question as it may appear. There is a widespread belief that when a new disease learns to spread among humans, it becomes less lethal. Many people believe there's an automatic trade-off between virulence—the severity of the disease—and transmission. Then a more transmissible variant of Covid-19 appeared in England in late 2020—and by January 2021, several studies suggested it was more virulent as well.[9] Suddenly health writers had to reconcile this with what they had always been told was "the evolutionary theory"—that, as one put it, "if it is too deadly, it will kill off its hosts. So if it starts to spread more, the lethality reduces, because if it didn't, there would be nobody left to infect." It shows how entrenched the idea has become, that this was expected to happen despite Covid-19's relatively low death rate up to that point and not the slightest sign it was killing people so fast it was running out of hosts.

This much we know: to survive, the pathogen needs to get into another host either before you die and take it with you, or before

234

your immune system kills it. So it can be good for the pathogen not to kill you all at once, if it needs you to stay alive long enough to spread its progeny, by coughing and sneezing and walking around if it's a respiratory virus. As a result, milder pathogens *may* win out over nastier ones as they adapt to humans, if the humans were otherwise getting too sick to spread them far. But this has been generalized into a broader conclusion: pathogens that are new to us are often severe at first because they haven't figured us out yet, but if they persist they *always* evolve to become milder so we will live long enough to spread them.

This is widely accepted as dogma. In 2005, near the peak of interest in pandemic preparedness, I spoke at a meeting on the topic at a very upscale British think tank and found myself in conversation about H5N1 bird flu with a then-member of COBRA, the top advisory committee called in by the British government in emergencies. I asked how he thought we could prepare for an H5N1 pandemic, if one emerged, given that it kills 60 percent of the people it infects. He looked me dead in the eye and said, "Don't worry. If it goes pandemic, that death rate will fall. These diseases always become milder." What, just like that? That's not usually how evolution works.

The 2018 UK flu pandemic plan, the only pandemic plan in the country when Covid hit, shows how entrenched this assumption has become. It is based, as a "reasonable worst case," on flu that kills 2.5 percent of cases, like the 1918 Spanish flu.[10] It acknowledges that H5N1 bird flu kills 60 percent, but simply states that if H5N1 becomes able to transmit between people, it will kill far fewer. While deaths above 2.5 percent "cannot be ruled out," it says, a pandemic H5N1 "would be expected" to kill around 2.5 percent. Moreover, it tells local authorities to plan for a 1 percent death rate, as, it states, a more lethal virus that attacks many people is simply unlikely.[11]

Recent history doesn't do much to support the idea that diseases from animals always become mild as they adapt to spreading in humans. Consider HIV. It spreads before you get sick, so it's no problem for the virus that after a few years, without the right drugs, 100 percent of infected people die. So there is no advantage for it to become less lethal—or as evolution scientists say, no selection pressure. As we have seen, it hit humanity in the early twentieth century, went very much pandemic, and despite some overoptimistic claims in 2014 that it was getting "less deadly," has shown no sign of getting nicer in all that time.

Or look at bird flu. It is a benign gut virus in ducks, as it really does need the duck to swim around pooping it out for a while if it's going to reach another duck, even after being diluted in pond water. But once it gets into a henhouse, it often mutates into a lethal, highly contagious infection. There is no shortage of chickens, so it doesn't need its host to stay alive a while to enable it to reach another one. Evolutionary success will go to the virus that gets into the next chicken and replicates before the others. And that is what happens: a simple mutation in the H protein, turning it into a virus dubbed highly pathogenic, allows it to replicate explosively and get into the next bird before other viruses. That this kills the chicken makes no difference to the virus. In fact, this mutation is beneficial enough for bird flu, at least in the short term, that this often happens when duck flu viruses of all kinds, not just H5N1, end up in a henhouse. In this case, getting a lot more deadly works out just fine for the virus.

Most worrying is that, as we saw in Chapter 5, a few mutations made H5N1 contagious between mammals, but it seemed to remain just as deadly. There didn't seem to be any trade-off between transmission and virulence. This is all the more worrying because another bird flu virus, H7N9, with a 30 percent death rate, had three of the five mutations required, occasionally spread a bit between people, and viruses with the highly pathogenic mutation as well could spread between ferrets and kill them, just by being inhaled. No apparent trade-off there.

Now, let me repeat that this does not apply, at the moment, to Covid-19. But neither is it true, as some scientists claimed early in the Covid-19 pandemic, that the SARS-2 virus is under no particular pressure to evolve. A typical comment: "At our cost, the virus is doing well enough colonising the human population. I don't see the drive for it to get nastier anytime soon."[12]

This is another widespread misconception, notes Andrew Read of Pennsylvania State University: viruses can be doing fine and still evolve. Random mutations constantly occur in these viruses, and if one just happens to enable the virus to spread even better in us, it should become more common in the virus population, or be "selected for" in evolution-speak, as spreading is what benefits viruses most.

Sure enough, by April 2020 a mutation in the spike protein SARS-2 uses to latch onto our cells, called D617G—code for the amino acids in the protein that changed and where they were in the protein—rapidly became the dominant form of the virus worldwide. And no wonder: research showed that "G," as it was nicknamed, replicated faster, making it perhaps 10 percent more transmissible than the original virus.[13]

Then in September 2020, England found a rapidly spreading variant of the virus called B.1.1.7, later named Alpha, with changes in no fewer than seventeen amino acids, which was around 50 percent more transmissible than G.[14] In November, as cases were skyrocketing in South Africa, the University of KwaZulu Natal found yet another, more-transmissible variant, later named Beta. In December, during one of the WHO's group calls for scientists worldwide—the kind modeled on the calls held during SARS—the South Africans alerted the British to look for one particular mutation in their new variant, N501Y. It was there. It had arisen completely independently, a strong indication that it gives the virus an advantage in spreading. Then another variant Gamma, in Brazil had it, too. By June 2021, the Delta variant from India was dominating infections in England and was apparently 60 percent more transmissible than Alpha and twice as severe. It also seemed to partly evade the immunity people had from vaccines or infection.[15]

This was predictable: all a virus needs is to transmit to another host and replicate, so any variant that does it better than the others will replicate faster and take over. It can also be an advantage for a mutant to evade the immunity induced in humans by prior infection, or vaccination, if there's otherwise enough immunity to slow the virus down. Any change in virulence may simply happen by accident—a virus with a particular mutation may transmit better by replicating faster, or binding more strongly to human cells (the Alpha variant does both those things), and that may happen to make the disease more severe. If more virulence doesn't slow the virus down, it will persist and, possibly, outcompete viruses without that mutation.

It's worth noting that increased transmission is enough to

worsen death tolls with no change in virulence. As epidemiolo-gist Adam Kucharski quickly pointed out when Alpha emerged, if a virus has a certain death rate and rate of transmission, so that there are, let's say, 129 deaths per 10,000 people a month, upping the severity 50 percent will boost that toll to 193. But boosting *transmission* 50 percent will mean far more than 50 percent more deaths, as the number of cases increases exponentially with every successive round of transmission. SARS-2 spreads at a rate that should give five successive rounds of transmission in that month. Exponential being what it is, this means around eight times more cases. With exactly the same death rate, now 978 will die.[16]

Of course, a more contagious virus that is also more virulent is the worst outcome—but in any case what emerges will be what-ever works best for the virus. A variant better at transmitting may happen to be less virulent, especially if, like flu in ducks, less vir-ulence is actually what makes it better at transmitting—the two really do trade off sometimes. But if becoming more transmissible happens to make a virus more deadly, and that makes no differ-ence to its success at transmitting and replicating—say, if it mostly spreads before you get sick—then that virus could well displace less contagious strains that are also less deadly.

Let's bust this myth: it is not a hard and fast rule of pathogen evo-lution that virulence always falls, or that there is always a trade-off between virulence and transmission. It is a bit worrying that people who are not primarily pathogen evolution experts, but are involved in pandemic response, apparently believe that it is. It is also worth remembering that Covid-19's waltz with humanity is just beginning.

Meanwhile, it's not just a question of how viruses act on their own: some vaccines can increase virulence. Andrew Read has done research with several diseases, including Marek's disease, a common plague of chicken farms, to see how vaccination affects the virus's evolution. He found that if a vaccine keeps the virus's host from getting sick, but still allows the virus to persist and spread—like the poultry vaccine for H5N1 did in China—it can select for a more virulent virus.

This is because while it is true that pathogens need not get milder in a new host, it is also true that they can be too deadly, kill off their hosts before they manage to spread, and die out. The occasional virus like that might emerge from random mutation, but it doesn't get far.

"Leaky" vaccines, though, induce immune reactions in the host that keep the virus at low levels, so an infected host doesn't get sick—but the virus doesn't completely die out either. If that happens, those viruses are free to become extra virulent because they don't have the problem of killing off their host if they do. And changes in the virus that increase virulence—faster replication or faster invasion of the host's cells, for example—might be helpful for a virus in a partly immune, vaccinated host, allowing it to persist and spread a bit better than the other viruses despite the host's immune reactions. Now if that virus reaches hosts who aren't vaccinated, it will be more lethal than usual. That is exactly what happens with chickens vaccinated for Marek's.[17]

This could be a concern if Covid-19 vaccines are "leaky," as the first few initially appeared to be—although in mid-2021 the Pfizer mRNA vaccine was found to block 94 percent of infection, making any leakage minimal.[18] "There certainly are plausible scenarios under which leaky vaccines could drive the [Covid-19] virus to increased virulence," Read told me. "I can also see scenarios where it could go in other directions." It all depends on what works for the virus. It is a good sign that Covid's relatives, the four coronaviruses that now cause 30 percent of our common colds, elicit immunity that stops the virus causing severe disease, but allows it to keep circulating and infecting people. That has suited these viruses just fine for centuries: they have never become more severe.

But every human infection is a chance for a virus to evolve, and the fewer chances there are, the less likely it will evolve at all. So as vaccinations began, the priority was, rightly, to get people vaccinated to reduce case numbers, leaky vaccines or not, while carefully watching for any changes in the virus, especially any that allowed it to escape vaccine-induced immunity. "Later, better vaccines will come along, hopefully," said Peter Hotez of Baylor University in Texas, a leading vaccine expert.

We certainly do not want a situation in which leaky vaccines, or anything else, foster much more virulent Covid-19. We are unlikely to vaccinate everyone—we never have with any other disease, even when we eradicated smallpox. If a virulent mutant of Covid-19 circulated silently among vaccinated people, then reached people who were not vaccinated, it could be bad.

Ultimately, though, Covid might settle down as a common, but mild, infection, if its cousins are anything to go by. In 1889–1890, a pandemic called the Russian flu started in central Asia and swept the globe, causing a million deaths—a teenage Winston Churchill wrote a rather bombastic poem about it.

Oddly, while deadly for older people, the 1890 pandemic did not affect the very young, completely unlike flu—but reminiscent of Covid. So was its tendency to be much more deadly in men than women. And the pandemic was followed by numerous reports of long-lasting symptoms of brain involvement, such as "brain fog" and chronic fatigue, again much like Covid-19.

We didn't know about viruses back then, and no one has looked for the germ responsible in tissue samples from the time. But Marc Van Ranst of the University of Leuven found in 2005 that a coronavirus that now causes the common cold, OC43, jumped to people from cattle in central Asia around 1890, and he suspects it caused the Russian "flu." Initially, as no human would have had any immunity to the new coronavirus, it would have caused a severe pandemic, just as Covid-19 did. But now OC43 is so common, most of us get it by the age of five, and as with Covid and the 1890 pandemic, as children we don't tend to get sick.[19]

So we all grow up with immunity to OC43 that prevents severe disease but allows it to keep circulating. In fact, this wouldn't happen if our immunity stopped the virus from circulating entirely, because then young children wouldn't get it from their elders and grow up with immunity that prevents severe disease. In February 2021 Jennie Lavine at Emory University and colleagues calculated that as long as we don't stop Covid circulating entirely, and our primary exposure is in childhood, the SARS-2 virus "may be no more virulent than the common cold."[20]

This scenario means that, like the 1918 flu, Covid will lose its terror once most humans are immune, either through vaccination or exposure, not because the virus "got milder." But viruses new to humans do adapt, and this can mean increased transmission and more death. The 2014 West African Ebola epidemic was by far the biggest ever, with nearly twenty-nine thousand known cases and around a 70 percent death rate. Yet there were only 315 cases in the best-known previous epidemic in an urban area, in the Congolese city of Kikwit in 1995. In the past, Ebola has actually been fairly difficult to catch, and outbreaks have been limited.

True to form, in 2014, the virus initially moved slowly from Guinea to Sierra Leone. Then something changed: it acquired a mutation in the part of the virus that latches on to human cells. After that, the virus spread much faster, reinvaded Guinea, raced across Sierra Leone and Liberia, and almost got loose in Nigeria, which stopped it using quarantine and the disease surveillance system built for the polio eradication program. After it acquired the mutation, nearly all the subsequent viruses in the epidemic had it—a bit like D617G in Covid-19. Jonathan Ball, a virologist at the University of Nottingham in the UK who tested the virus, says it was almost certainly an adaptation that allowed it to spread better in people, just as you'd expect. But adapting to us did not make it milder.

Jeremy Luban of the University of Massachusetts, who also studied the virus, agrees that the mutation showed all the signs of being an adaptation to people: it emerged after unprecedented circulation in people, allowed the virus to bind better to human cells, and then dominated the rest of the epidemic.[21] Andrew Read observes that Ebola spreads when severe cases—and recent corpses—shed the virus in body fluids. So becoming less deadly is unlikely to be an evolutionary advantage for an Ebola virus. The mutant should have disappeared with the stamping out of the 2014 epidemic but could emerge again as viruses lurking covertly in organs like the eyes or testes of people infected in 2014 cause resurgent infections, like one in Guinea in 2021.[22]

How did the belief that pathogens always become nicer when they start spreading more easily become so widespread? It started, says Ed Feil of the University of Bath, when scientists in the late 1800s noticed that diseases of cattle became milder after a herd had repeated outbreaks and, knowing little about the cattle's immune systems, assumed the virus was adapting.[23] But, says Read, the idea later got a big boost from myxomatosis in rabbits. That virus causes a mild disease in animals in the rabbit family in its native South America, but it causes a lethal illness in European rabbits. It was released in 1950 in the Murray Valley of Australia to control European rabbits, which had become an invasive pest. The same thing was done by landowners to cut rabbit numbers in France in 1952 and the UK in 1953.

The story you hear is that the virus rapidly evolved to become a mild disease, as the original strain killed too many rabbits too fast, ran out of nearby hosts, and failed to spread, while milder viruses thrived. This is the story people cite when they say a virus that learns to spread well in us will become milder.

In 2015, Read and his colleagues did a review of the research into what actually happened. A few months after its initial release, the virus exploded across southeastern Australia. The toll was incredible: it killed some 95 percent of rabbits in farming areas. Much the same thing happened in Europe: the British children's classic *Watership Down*, about a band of rabbits, called it "the white blindness."

A year or two later, Australian virologists started finding slightly less lethal viruses among surviving rabbits. These were not, by any standard, mild. The most common killed 70 to 90 percent of laboratory rabbits, which are the same species as the wild ones. But it took longer to do it than the original strain, so it did give the virus more chance to spread. The researchers also found strains that killed fewer than half the rabbits, but they didn't transmit well, as the rabbit's immune system tended to kill them first, so these were rare. In a natural experiment rare at such a large scale, the release of the virus in similarly nonimmune rabbits in Europe followed the same course.

The die-off put a massive "selection pressure" on the wild rabbits. Basically, the few remaining rabbits survived because they had genes, probably for particular kinds of immune reaction, that made them more resistant to the virus. While the virus persisted, having those genes was a real advantage, and rabbits with them rapidly became the majority. Being rabbits, they replenished the population, so the new ones were resistant to myxomatosis: seven years later, the virus was killing only a quarter of the wild rabbits. But it was as lethal as ever in lab rabbits, which hadn't been selected for resistance.[24]

So the virus did evolve to become somewhat slower to kill, but not mild—and the disease overall became less lethal, not because the virus did, but because the few surviving rabbits gave rise to a population that resisted it.

Then in 2017, Read and his colleagues discovered the virus was fighting back: now it gets around the rabbits' resistance by attacking the immune system directly and spreading more readily. So again, adapting to its host meant upping transmission—and becoming more lethal.

Scientists who are actually specialists in this area don't talk about viruses becoming mild so they can spread, as though the virus somehow surveyed the situation and decided it had best do that. They talk about an arms race between the virus and the host. Myxomatosis did become a bit less lethal, but that was only after it had killed nearly all the rabbits. There was no pressure for it to become any milder until that happened, and even then it was still pretty lethal. The disease overall looked milder because the surviving rabbits were resistant. Then the virus got worse again.

Somehow, I don't think that was what that COBRA guy had in mind when he told me not to worry about H5N1 because new viruses always get milder. I hope the people doing our pandemic planning are thinking more carefully, and knowledgeably, about what really happens when viruses adapt to us.

If we are hit by a much more virulent virus, how bad could things get? We have established that our globalized, interdependent world is surprisingly fragile. The domino effects of a pandemic on global production and trade can severely damage an urban economy dependent on just-in-time goods and services from the rest of the planet. That much we are learning from Covid-19.

But a more virulent pathogen means you lose people—not temporarily to lockdowns, but permanently. What is the impact of that? We are losing people now, of course, but on nothing like the scale we could from a really bad pathogen—and SARS showed us that a virus with a much higher death rate than Covid-19 can get very close to pandemic. Beyond the immediate tragedy and sorrow, what is the impact on a complex, fragile world like ours of a lot of us dying?

It isn't necessarily self-evident. A lot of our problems, as we have seen, stem from the way we are managing our unprecedentedly huge numbers, including the pressures of poverty and economic competition that lead to our encountering new pathogens. A friend of mine was once listening to me going on about bird flu and pandemic threats and said, "Look I don't want to sound callous or anything, but, well, wouldn't it be better in some ways if there were fewer of us?" That was the question I set out to answer.

What I found was that our globalized, industrial society is effectively arranged vertically, with almost everyone totally dependent on support by many other human subsystems called critical infrastructure: housing and heating, food production and distribution, water supply and sewage management, public health, transport systems, security services, telecoms, banking, shops supplying essential goods and services, electric power. To some extent, all the subsystems rely on each other. With all these complex interdependencies, this makes us, basically, a big game of Jenga: pull out a few pieces, and the rest can fall.

What may not be immediately apparent is that a lot of the most important pieces are the people themselves, says Yaneer Bar-Yam, head of the New England Complex Systems Institute. It isn't obvious, he told me, but research with complex systems shows that the more complexity rises, the more individuals matter. If a more lethal pandemic took out more of the key people running our critical systems, the impact could be pervasive.

Some industries are hubs—like bats in an ecosystem, a lot of other parts depend on them. Industries in turn depend on their workers. In 2000, a strike by truck drivers blocked nearly all gasoline deliveries from Britain's oil refineries for ten days. Public transport collapsed, grocery stores emptied, hospitals ran minimal services, hazardous waste piled up, bodies went unburied. The government had to step in. A subsequent study predicted economic collapse in Britain if all road haulage, not just fuel deliveries, was shut down for only a week.[25]

Today, we all depend even more on just-in-time deliveries: if the trucks stop because drivers are locked down, or sick, or dead, or caring for sick family, cities will rapidly have no food, vehicles won't have fuel, food in depots will rot. Making more such deliveries with automated transport systems in the future may seem to fix this, but the truck driver is only one link in a complex web of distribution. Unless every aspect of food and fuel production and distribution is automated to the point that none depends on people—and this seems unlikely for the foreseeable future—then if certain hub industries are paralyzed by the loss of people, the impact can be devastating. Many other choke points depend on people: doctors and nurses, engineers who run power grids or essential manufacturing, and global supply chain managers are not all readily replaced. Even transient absences of key workers can cause snowballing problems. During Covid-19 lockdowns, oil refineries shut due to plummeting demand as air and road traffic fell. In a pandemic with a higher loss of people, absence of workers at oil refineries starts becoming a problem.[26] The current UK pandemic guidance for the natural gas industry predicts that anything more than staff absences up to 30 percent for a month "would be problematic," whereas an absence rate of 45 percent—or possibly less during peak demand in winter—could trigger a Gas Deficit Emergency, with some users, like factories and homes, shut down.[27,28]

As always in a complex system, the problem doesn't stop there.

Oil refineries produce transport fuel, and lack of that stops deliveries, including of coal, on which much electricity in some countries still depends. That's where things really start to collapse, Michael Osterholm told me. Failure of electricity will cripple subsystems from lighting to ATMs to refrigeration to pumping drinking water, and electric power is needed for mining coal or pumping oil for generating the electricity itself. It gets worse from there.

Truck drivers and refineries are only two sectors where this domino effect could start. Once one part of a network of interdependencies wobbles, the rest is at risk.

As I write, the Covid-19 pandemic does not have a very high rate of sickness and death, so large percentages of the population are not being disabled by disease. But the social distancing measures being used to slow its spread are stopping much economic activity. Tellingly, workers deemed essential to critical infrastructure are exempted from this everywhere. There are some cogs in our system that we really need to keep turning. Tragically, those "key workers" are at increased risk of infection and death.

The official US list of essential workers for this pandemic makes fascinating reading.[29] A random selection: health care workers at all levels, including cleaning staff; building security staff; food workers; crop pickers; miners; armored cash transporters and ATM servicers; powerline repair people; truck stop operators; grocery store workers; the people who cut tree branches away from overhead electrical lines; sewage processing plant workers; road repair crews; bus drivers; plumbers; waste disposers; telecommunications repair people; IT workers who maintain the internet; metalworkers; chemical workers; laundromat staff; janitors . . .

Of course, there are also judges and lawyers, doctors and power plant engineers, cyber-defense experts, some clergy, and other white-collar workers on the list, but a glance shows that a lot of critical infrastructure depends on low-income people. It has long been known that low-income people are more likely to die of infectious disease generally, due to more underlying poor health, in turn due to greater exposure to environmental stress and toxicity, poorer diet, poor housing, and lack of access to medical care.

More relevant to Covid, many poor people are key workers, stacking supermarket shelves, staffing checkouts and fast-food takeout, cleaning hospitals, harvesting crops, or caring for the elderly, work which must continue and cannot be done from home. Data from 45 million mobile phones in twelve major cities across the US during 2020 showed poorer people were less likely to stay home during stay-at-home orders, making avoidance of infection "a luxury" beyond their means.[30] The poor also use more public transportation, exposing them to more potentially contagious people; are less likely to have sick leave; and are more likely to have a home that is too crowded to allow social distancing and includes several generations, exposing vulnerable older people to working younger ones. An early British study of patients hospitalized with Covid, released in May 2020, found those who died were almost two times more likely to come from a poorer neighborhood than a richer one, and it was only partly because they had more underlying illnesses like diabetes.[31]

It might have been, in part, because they were more likely to be people of color. Repeated studies in several countries have found racial minorities are more likely to die of Covid. They have more of the underlying conditions that worsen the disease, but this alone doesn't account for the increased deaths. Several studies have found that, due to housing or jobs, people in racial minorities are simply more likely to be infected. Even in Nordic countries, which do not have a long history of immigrant racial minorities, Somali people were more than ten times more likely than the average resident to be infected.[32]

Poverty doesn't account for all of it, though it doesn't help. Samrachana Adhikari and colleagues at New York University compared rates of Covid-19 infections and related deaths in richer and poorer urban counties in the United States that were predominantly nonwhite—of African or Hispanic descent—with counties that were predominantly white. Richer nonwhites were nearly three times more likely both to get infected and to die than richer whites. Astonishingly, poorer nonwhites were 7.8 times more likely to be infected, and 9.3 times more likely to die, than poorer whites. Being Black seemed more deadly than being poor.[33]

Similarly, in Britain, people of African and South Asian descent are three to four times more likely to die of Covid than whites of the same age—and minority children up to six times more likely to get the relatively rare, sometimes deadly pediatric inflammatory syndrome.[34] Underlying illness and where and how you live accounted for less than half the difference, reports England's Office for National Statistics.[35] Daniel Ayoubkhani of the ONS says being a key worker, or having one in your household, was a risk on its own.

The cause of the rest of the increased risk wasn't clear—but it largely disappeared during lockdown, so it seems likely to have been due to increased risk of infection rather than a greater innate tendency to get severe disease. Independent studies also found Black and South Asian people in Britain were more likely to be infected but not more likely to develop severe disease once they were.[36]

So this isn't biological. Adhikari's team found that, in New York City, "Black and Hispanic populations are not inherently susceptible to having poor Covid-19 outcomes. If they make it to hospital they fare as well or better than their white counterparts."[37] Many, however, did not make it. Structural racism, a stew of factors from poorer health communication to de facto segregation in workplaces and in housing, could account for this.

The world is trending toward greater inequality, especially economic, but with rising migration, also racial. In richer countries, which have significant immigration, that means more people who are extra vulnerable in a pandemic—and many of them are the people responsible for critical infrastructure. Some—for example, in the meat-packing industry, where social distancing has been absent and Covid-19 has struck hard, but also in health care and many other sectors—are recent immigrants, some of whom are undocumented and have less access to health care and sick leave.

Greater vulnerability among lower-income people worsens the spread and impact of a pandemic in the most critical parts of our complex system: firefighters, paramedics, police, care workers, the people who produce everyone's food, drinking water, electric power—the list goes on. The less those people can withstand a pandemic, the more the system that supports everyone is at risk of collapse. More inequality, more poverty, even more racism, means more risk.

No pandemic plans that I have been able to find seem to take into account the domino effects propagated through our complex critical systems simply by the deaths of critical people within them.

Most engage in wishful thinking about death rates. The UK flu pandemic plan with its assumption of less than half the death rate of 1918 is typical; Tim Sly, an epidemiologist at Ryerson University in Toronto, says he has never found one that assumes an even worse death rate than that of the 1918 flu, even though we know there are deadlier flu viruses, never mind other kinds. Perhaps planners assume that if viruses go pandemic, they will become mild; perhaps the alternative is simply too harrowing to consider.

Tabletop simulations of pandemics with real-life government officials and industry bosses find many aspects of society rapidly collapse as unexpected consequences pile up—and, I was told by people who conduct such simulations, participants always discover, to their surprise, that their key personnel actually are the critical infrastructure. If a highly lethal pandemic takes out a lot of people, the consequential failure of our support systems may go on to take out more.

We rarely think about just how precarious these systems are, but the evidence is everywhere. A common saying in security circles is that a city is never more than three or four meals away from anarchy: as food prices on the global market rose in the run-up to the financial crisis of 2008, there was rioting in many places. And systems support each other: for example, if chlorine for water purification cannot be delivered at the same time that power outages make it hard for many to boil water, waterborne diseases could result. Often we don't see the problem until it's in our faces: New York discovered after Hurricane Sandy that high-rise residences dependent on elevators became traps for people less able to handle many flights of stairs when the power went out, while even hospitals with generators failed to cope.

Countries rely on foreign deliveries for everything from milk cartons to pharmaceuticals, a lot of them delivered by sea. In a bad pandemic, shipping will falter even more than it has with Covid-19. Even in this one, large oceangoing vessels, be they cruise ships or aircraft carriers, were at increased risk for contagion, and hundreds of thousands of crewmembers were stranded at sea or in ports from which travel restrictions stopped them getting home.[38] Container ships are nowhere near as heavily crewed or luxurious as the cruise liners that made headlines early in the pandemic, but they still need people—and once losing people means losing shipping, virtually all industries will be affected.

A pandemic with a high death rate could trigger a lot of this kind of domino effect. Another thing about complex systems is that they lose complexity rapidly but gain it back with difficulty, if at all. Partly, this is thermodynamics: the first process releases energy so can happen spontaneously, whereas the second requires it and needs help. But in addition, as we mentioned, the study of complex systems has revealed that these systems settle in stable states, from which it is hard to dislodge them.

And that brings us to collapse. Collapse is tipping from one stable state to another with less complexity, providing fewer services and able to support fewer people. If the collapse of various subsystems of our current society propagated globally, sweeping up other subsystems in the process, eventually some countries, industries, or economies could collapse as well and have trouble getting back up. The more this encompasses people and processes essential for life support, the more likely the collapse is to be existential, for some or all of us. Such a dramatic event might seem unimaginable, but Covid-19, even though it is, let us repeat, not nearly that drastic, has laid bare just how interconnected and fragile some of our systems are. UN Secretary-General António Guterres called the pandemic a wake-up call. "We have an opportunity now to do things differently," he told the BBC. "It is clear the world is too fragile in relation to the global challenges we face. That fragility was demonstrated obviously with the pandemic." It will be even worse, he says, with climate change.[39]

On the bright side, an understanding that we all depend on complex systems can help us prepare for some of the challenges Guterres mentions, including the next pandemic. That means making our systems less fragile—but the right answers may not always be the most obvious. Over the past few decades, many jobs in the traditional industrial heartlands of Europe and North America have been lost to "offshoring," the movement of industries to fast-industrializing countries elsewhere. There is now talk of rolling that back and "reshoring" in some industries, especially those that are vital in a pandemic, to shorten fragile supply chains.

But that might not always be the best thing to do. In February 2021, the European Commission found that reshoring efforts in the UK, the United States, and Japan have had only "modest" success, and recommended it only for critical sectors—such as the essential, generic drugs now almost solely supplied by China.[40]

Shannon O'Neill of the Council on Foreign Relations, a US think tank, warns that in many cases, reshoring will lose offshore industries the advantages of scale and labor costs they gained by moving, resulting in rising costs of goods for ordinary people, by some $10,000 a year on average for US consumers, a significant portion of average incomes.[41] This would mean hardship for some.

The disappearance of those industries from the largely developing countries where they are now would mean hardship for many more.

Moreover, it could be prohibitively difficult to reassemble a complex system like manufacturing in a new place: O'Neill cites an effort by Apple to make MacBook Pros entirely in the United States in 2013, which failed because one type of screw could not be sourced locally. It almost sounds like the old saying: for want of a nail the horseshoe was lost, for want of the horseshoe the horse was lost, then the rider was lost, then the message, the battle, and so on, until the kingdom is lost. Perhaps we have always had an instinctive understanding of complex systems—though it often seems that few understand our society's fragility.

Some just don't want to. Part of this chapter on complexity was posted on *Medium* in mid-2020, and a vociferous faction of commenters insisted it was wrong, because our society simply is not a complex system. Complicated, they agreed—but not complex, with all the nonlinear properties and fragilities that implies. No expert in complex systems I know of agrees. It's hard to understand how someone can believe this assertion, although perhaps not hard to understand why one might wish to.

Many more analysts do not so much deny the implications of complexity as simply overlook them. They argue that, because natural disease has never wiped out humanity throughout our past, it is unlikely to in future. Some allow that novel disease germs created in labs are an unknown quantity, so one of them might—but on past performance, natural disease seems unlikely to pose an existential risk.

This overlooks the fact that society as a whole is far more vulnerable to disease now than in the past, for all our science. In 1347, the Black Death killed as much as half of Europe's people, but the survivors, and their relatively simple, agrarian society, staggered on. Now losing key workers or supply chains in a highly specialized, complex industrial society would cause cascading failure in our support systems, killing people spared by the disease itself.

And before the 1800s, an outbreak in one part of the world wouldn't reach the rest; now a new virus circles our globe in days. And in the past when a civilization like Rome or the Maya collapsed, there were others—China, Cahokia—that would continue unaffected. Now just one, fragile, interdependent civilization spans the globe. A bad pandemic, natural or not, might destabilize it enough to unleash further disaster—and nuclear conflict could be terminal. Then because we need energy to relaunch our vital energy-generating systems, it isn't obvious we could reboot if they went down.[42]

But while complexity increases risks, understanding it suggests solutions. O'Neill suggests putting more redundancy back into globalized industries, making them more resilient. It will cost, as efficiency was adopted in the first place to save money. But it may well be less expensive than dealing with collapsing supply systems in the next pandemic. Shipping analysts were already saying in April 2020 that they expect industries to now diversify their suppliers— even if that means higher costs.[43]

Homer-Dixon agrees that we need more redundancy in the system, but also less overall complexity, to increase the "slack" in our support networks that can absorb shocks. That could, he says, mean cutting international travel, simplifying global supply chains, and indeed bringing some crucial production closer to the final users, or at least putting it in more places.

He says it isn't just our connectivity that puts us at risk, but also our uniformity, not just biologically as humans but culturally, in our food, ideologies, social media, finances, consumerism, even our antibiotics. If we have the same responses to perturbations everywhere, we risk disaster everywhere if one goes wrong. "Diversity, often a key feature of complexity, can be highly beneficial," he told me. The problem is not complexity per se—it's whether it leaves you more, or less, vulnerable.

Too much uniformity sets us up for cascading failure—and also synchronized failure of apparently independent subsystems. Homer-Dixon led a group of leading complexity experts that reported in 2015 that the apparently separate crises of 2008–2009, when both food and oil prices rocketed just as a US mortgage crisis triggered financial turmoil, were deeply linked.

The global economy has never gone back to the way it was before the crisis and has apparently found an alternate stable state.[44] Homer-Dixon predicts that this pandemic will similarly be "a global tipping event, in which multiple social systems flip simultaneously to a distinctly new state." And, he says, if we don't start managing the problems raised by our complexity, we will get more of them, with ever-higher destructive force. The potential problems of a bad pandemic pale beside some of the possible impacts of climate change, with large swathes of the globe becoming too hot to live in, coastal megacities drowning, and vital crops failing.

A pandemic of a far worse disease than Covid-19 could be

another impact of warming, as novel infections are released by upended ecosystems and uprooted human communities and support systems. Deaths of large numbers of people would be hardship enough—but it would have the further insidious effect of taking out many of the key lynchpins in our complex global systems, entailing further losses. "No" is the answer to the question my friend asked: we can't lose a lot of people and put less pressure on the planet while things continue as before. We would lose a lot of people, then a lot of everyone else, and nothing would continue as before.

So what would be the result of facing something like the Black Death in our modern society? We are not as resilient as Europe in 1347. The result could be the generalized collapse that has overtaken every earlier civilization. These are always accompanied, Tainter told me, by steep losses of technology, knowledge, and people. He is dubious of our ability to deliberately step down our complexity to stave that off. If you needed another reason to stop the next pandemic from happening, consider that.

All this reminds me of a tale told by the author Douglas Adams in one of his Hitchhiker's Guide books, *The Restaurant at the End of the Universe*. The planet Golgafrincham had too many people. So it contrived to keep the top professionals and the low-level practical workers, but rocketed all the middle-level "useless" people into space: security guards, for example, and telephone sanitizers. The remaining population subsequently lived happily—until they all died from a virulent disease contracted from an unsanitized telephone.

# CHAPTER 8

# The Pandemic That Never Should Have Happened — and How to Stop the Next One

> We've got to dance with the virus.
> There's no choice.
> —George Gao, head, China Center for
> Disease Control and Prevention[1]

In two campaign speeches while running for president, John F. Kennedy said, "When written in Chinese, the word 'crisis' is composed of two characters—one represents danger and one represents opportunity."[2] Kennedy popularized the notion, and its use became widespread, including an appearance in Al Gore's Nobel Prize acceptance speech.

However, while it may have made for a good speech, it is not true. Apparently, this idea was the result of an optimistic mistranslation by Western missionaries in China in the 1930s. In fact, the first character does mean danger, but the second one just means a time when things happen or change.

Covid-19 has been, by anyone's reckoning, a crisis—and its impacts are just getting started. Things are going to happen or change now, whether people use the opportunity to take control of them in the broad interests of humanity—or not. It may be a chance to achieve things we could not achieve before, as summed up in the popular slogan "Build back better." The popularity of Kennedy's statement shows that we recognize this deeper truth—that crises can provide those opportunities. Or we might just be swept along by the economic and political storms the pandemic has unleashed and never deal with any of the underlying problems that got us here.

That would be tragic. In a minute, we'll look at our options. But first, let's look back at where we started and how we got to this point, where we can look toward the future. I called this the pandemic that never should have happened—and said we could possibly stop something like this from happening again. Let's see how that adds up.

In Chapter 1, we saw that Covid-19 started as a cluster of unexplained cases of pneumonia in Wuhan, China, which was recognized in December 2019, although there were possibly unrecognized cases in early October.[3] In late December, Chinese authorities told the WHO about it—but also said the virus did not spread person to person, even though the doctors knew it did. With that as the official story, no large-scale containment efforts and public health messaging aimed at slowing spread of a contagious disease could be undertaken in Wuhan.

It's hard to imagine what health officials, who ordered doctors to keep quiet as the epidemic grew, thought would happen. Maybe they thought they could keep most cases of the infection safely contained in hospitals. Everyone remembered SARS, another coronavirus, and people with SARS normally didn't spread the virus until they were quite ill. Also, there were no cases of SARS with few or no symptoms that were nonetheless contagious, as there are with Covid.

Secrecy can also simply become an instinct in authoritarian systems, says writer and sociologist Zeynep Tufekci.[4] As we saw in the chapter on SARS, China deemed outbreaks state secrets unless officials gave permission for them to be made public. This is not unique or new: the International Health Regulations, now a cornerstone of global epidemic management, arose from efforts in the 1800s to stop governments from keeping cholera outbreaks secret and causing problems for shipping. But in China secrecy not only hampered the early response to Covid, it amounted to repression: in December 2020 independent journalist Zhang Zhan was sentenced to four years in prison for sharing videos of the early lockdown in Wuhan, another named Chen Qiushi reportedly went missing for months, and others allegedly remain missing.[5]

In early January 2020, officials in Wuhan obscured matters further by decreeing that someone could be tested for the new coronavirus only if they had been exposed either to the now-closed fish and wild animal market linked to many early cases or to a known case. As the virus spread in the population, increasing numbers of infections had no such links, so this guaranteed they would not test many cases. But let us recall that Europe did the same thing initially with swine flu in 2009, and early in the pandemic many places, including states in the United States, refused to test people for Covid-19 if they had no direct link to China, even after the virus was known to be elsewhere.

In China, Tufekci suspects that a culture of suppressing bad news and passing decision-making up the hierarchy may have meant that President Xi Jinping didn't know how bad things were in early January, even though senior health officials did—but in an authoritarian system, only he could change the story. Things were bad, though: by January 20, there were so many cases in Wuhan and increasingly elsewhere that only drastic containment measures ahead of the Lunar New Year holiday could prevent the virus spreading out of control across China.[6] Chinese scientists announced the virus was contagious, and the lockdowns began.

As we saw, research suggests that if those measures had been taken earlier, the epidemic might have been knocked back, although perhaps not knocked out. But would anyone have realized those measures were needed? The virus that causes Covid-19, we now know, can be difficult to stop by just isolating cases and tracing their contacts without added social distancing, even though that worked for SARS, as China's scientists would have known. Covid-19 is much easier to catch than SARS, and unlike SARS, it is spread by people without symptoms. The measures imposed after January 20 meant cities in China outside Hubei province, where Wuhan is located, never needed the total lockdown used in Hubei, but as the WHO reported, many of them found they needed social distancing as well as isolation of cases and contact tracing to halt the epidemic. In early January, public health officials in Wuhan didn't know any of that.

Mathematical modeling shows that the kind of measures China eventually took in late January, including severe limitations on people's movements, would have vastly reduced the size of the epidemic if they had been applied in early January. But even if authorities had gone public about the virus being contagious then, it isn't clear that they would have imposed control measures that extreme, knowing only what they knew about Covid-19 at the time. They would probably have done what worked for SARS, and it would not have been enough. Besides social distancing, widespread testing to catch presymptomatic or symptomless cases would also have been needed. It should be noted that, even knowing a lot more about the virus than China did at the outset, several countries, including the United States and the UK, were slow to impose the testing, distancing, and containment required, and that did work—as demonstrated by countries that did impose these measures, like South Korea and New Zealand.

So it seems unlikely Covid-19 would have been stopped completely if China had publicized the full situation and taken more extensive public control measures earlier. But the spread of the virus, within China and out of it, might have been more limited and more controllable.

And if scientists and health authorities everywhere could have used this head start to tackle Covid-19 earlier and used viral sequences from China to test travelers from affected areas and look for cases at home, we might have headed off the eventual steep rise of infection in more places. But the virus still would have invaded poorer or less controlled countries that might not have taken these measures, and would have multiplied there, making it hard to prevent global spread. We would have needed an earlier massive shutdown in air travel to prevent that, which would have been unlikely.

Certainly there are what-if scenarios that suggest much earlier containment measures in China could have meant no pandemic at all, or maybe an epidemic in China and a few controlled outbreaks around the world. But the real clincher for me is looking at how many countries disregarded the WHO's advice on control measures, even after it was clear how bad the virus could be. Even if they'd known earlier, I'm not sure how many would have done what was needed in time. Hindsight helps you win the next battle, not the last one.

Of course, the whole point of this analysis is to see if, and how, we might do better next time. And it suggests that, because of this battle, we can at least hope that next time countries won't be as slow to see the danger they are in from a viral infection emerging anywhere. With luck, we have at least been jolted out of the blind complacency and outright denial about infectious disease that delayed most of the world's response to Covid-19. But to reap those benefits we need some agency empowered to spot the danger and drive the global response.

So, *lesson for the future number 1*: we need a high-level, author- itative system bringing countries and international agencies together to collaborate on disease, in peacetime as well as emergencies, so that no one conceals important details about worrying outbreaks, and everyone works together from the beginning. At the very least, we need surveillance systems that will spot clusters of cases early, when an infectious pathogen might still be contained—that com- puterized alert system China installed after SARS, or something like it, in far more places would be a great start, especially if the alerts were shared widely. We will look at possible ways of doing that later.

Also, the world needs to start taking the threat of pandemics, and the warnings of its scientists, seriously. Covid-19 is taking care of the first part of that. As for scientists, that will always depend on how inconvenient their advice is versus how venal their govern- ment is. But we can at least hope listening to scientists will become more of a norm now that Covid-19 has shown us how desperately a modern society needs to rely on facts, evidence, and honesty, rather than secrecy, ideology, or wishful thinking.

Having looked in Chapter 1 at how Covid-19 emerged, we then in Chapter 2 looked at emerging diseases generally. By the 1960s, much of the world had largely defeated the old infectious diseases with prosperity and vaccines, whereupon we disinvested in the kind of public health needed for infectious disease, despite the wake-up call of AIDS in the 1980s, warnings of more new diseases from US scientists in 1992, and evidence by 2008 that we were contracting zoonoses from wildlife at an increasing rate. The WHO made a list of the most worrying pathogens, including coronaviruses and horrors like Ebola and Nipah, so we could make vaccines and diagnostic tests for them. Very few are ready yet.

I call Covid-19 the pandemic that never should have happened. Yet scientists have been warning, increasingly, about the growing risk of pandemics since 1992. How is it possible to warn that something is going to happen and then say it never should have happened?

Easy: that is the whole point of warning. We didn't act enough on the warnings, and there have been plenty. I wrote an article in 1995 titled "Can We Afford Not to Track Deadly Viruses?" It was about a WHO plan to monitor emerging diseases in the wake of an Ebola outbreak in Central Africa—except WHO member states weren't going to approve enough funding.[7] Could we have done better? We could certainly have improved our systems both for spotting emerging diseases and for responding to them. The willingness of countries to pay for that kind of surveillance and response increased somewhat after 1995, but not enough.

*Lesson for the future number 2*: Now that we know the warnings were real and the cost of disregarding them is great, maybe there will be more willingness to improve our systems for prevention monitoring and response, first by curtailing activities that increase the risk that diseases will emerge, then by beefing up surveillance for outbreaks, and finally by investing in drugs, vaccines, and diagnostics for the threats we already know exist. Let's not let the coronavirus take our eye off Nipah and the others. Especially Nipah.

Coronaviruses were on the WHO's list of worrying viruses because of what Covid-19's relative did seventeen years before Covid appeared. Talk about being warned. In Chapter 3, we looked at three warning shots from previous outbreaks of coronaviruses: SARS, MERS, and, in pigs, SADS. SARS offered us two big lessons that nations clearly haven't learned: protect health care workers and immediately tell the world everything when a threatening new infection breaks out. There has been progress on that second lesson since 2003, but obviously not enough, given what happened with Covid-19 in China. Our defenses against these viruses have been stymied by problems of capitalism as well. Despite all the warnings, we didn't develop any remedies for coronaviruses because after SARS was stamped out there wasn't an obvious market for them. Producing medicine for the public good as well as profit may be coming back, and it's about time.

*Lesson for the future number 3*: There are two parts to this lesson. One is about laying in supplies of existing remedies. We need PPE, personal protective equipment, for health care workers—SARS should have taught us that, but now Covid-19 is forcibly reminding us. Besides serious stockpiles, we need surge capacity in manufacturing. If countries don't learn even that much after the toll

Covid-19 has taken on nurses and doctors in so many countries, I despair. We all should.

The second is about developing new ones. Profit-driven markets can do wonderful things, but not everything. We need to stop relying on them to do what only governments can do and develop products that cannot immediately turn a profit, but that we nonetheless need for the utterly essential public good, including new antibiotics, vaccines that everyone can afford—and better ventilators, because respiratory viruses will always be among our biggest threats. The United States tried to do that for Covid and failed, another time market forces trumped public good when they shouldn't have.[8] There are dozens of calculations showing the cost equivalent in fighter jets or nukes, which governments apparently can afford, compared with the costs to develop, produce, and stockpile the life-saving medical goods we need. The resources are there.

So, what about stopping these viruses at the source, or at least knowing enough about the sources to know what's coming? In Chapter 4, we looked at bats, why they have so many viruses, why killing them is a very bad idea, and how the Wuhan Institute of Virology not only found the exact viral gene sequences of SARS lurking in bats from one cave, but also found viruses that were very close to the one that later caused Covid-19. Meanwhile, both they and a lab in the United States found that these viruses, straight from the bat, caused disease in mice primed with the human receptor protein ACE2 and had no trouble invading human cells.

In the scientific papers they published, the researchers issued very explicit warnings about the pandemic potential of these viruses, far more urgent and substantiated than for many other viruses of pandemic concern. There appears to have been no action taken on these warnings, except that the US government research project that collaborated with the Wuhan Institute had its grant renewed—only to have it canceled again when, amid the hysteria of the pandemic, unsubstantiated allegations arose that Covid-19 actually escaped from the labs that tried to warn us about it.

Outside commentators with various ideological axes to grind have seized on the labs' warnings to blame the labs themselves for the pandemic, in a kind of molecular version of shooting the messenger. It's worth noting that those labs did this work for fifteen years and more without any sign of a problem. Meanwhile, the same species of bat that was found carrying similar viruses lives across Asia and Africa, there seems to have been a colony of live bats in central Wuhan, and bat feces is a widely used Chinese traditional eye medicine. Surely all that seems a greater risk.

For now, *lesson for the future number 4*: when publicity-shy, certainty-averse scientists put these traits aside and start screaming that there's a really threatening thing out there, we need to listen and make it someone's job to respond. I have no doubt this lesson will eventually filter through, when climate change starts causing massive crop failures, uninhabitable cities, and unprecedented waves of refugees. By then, of course, it might be a little late to act on it.

But here it is, the ultimate reason why Covid-19 is the pandemic that never should have happened. We may or may not have been able to contain it once it jumped from bats to humans—*but it should never have jumped*. We knew enough fifteen years ago to begin avoiding bats, and bat products, and bat anything that might transmit their wealth of viruses. And according to all the science we know now, the Covid-19 virus came from bats—not civets, not pangolins, not raccoon dogs, and certainly not snakes (that early claim was never anything like scientifically valid).[9] It came from bats, and so do a lot of other viruses.

But we need bats because the rest of our ecosystems depend on them, especially globally vital resources like rainforests, never mind our food crops. So, we should give bats plenty of space. We certainly shouldn't build open livestock pens under their roosts, and perhaps we should give people who can't avoid encountering bats in their environment extra disease surveillance and health care to swiftly catch any virus that jumps. But it's actually not easy to catch a virus from a bat. We really had to work at this, by destroying or invading bat habitat and using products made from them. Let's stop.

In connection with this, may I respectfully make a suggestion? Tradition is very important, and traditional medicine is often valuable, but perhaps using bat feces to treat eye disorders is one practice we might consider letting go of. This is not because it's feces—indeed, Western medicine is only now learning uses for that long known in China—but because of what we now know about bat viruses. It is admirable that the remedy was taken out of the *Chinese Pharmacopoeia* in 2020, but until it is clearly made illegal and that ban is enforced, it is unlikely to disappear from Chinese traditional medicine shops. I'm not sure why people assume that for Covid-19 to emerge, a bat virus had to be transmitted via an "intermediate" species or a laboratory escape, when a lot of people are using bat feces as eye medicine. Granted, many of the viruses present in the feces may die as it is dried, but do all of them, every time? It would take only a few surviving viruses, and the bad luck that they happened to be among those that take to humans, to unleash another in us. And even if, improbably, drying the stuff means users are never at risk, the people who gather and process it are, and they transmit viruses like anyone else. The fact that at least some purveyors of traditional Chinese medicine online had stopped selling bat feces by May 2020 "because of Covid-19" suggests the risk is being recognized—but it could take a concerted effort to really get it off the market.

The people trying to at least clean up live animal, bushmeat, and wildlife markets everywhere, not just in China, also have a point. There is little evidence that Covid-19 in fact started in a wet market, but SARS probably did, and we know other viruses lurk there, notably bird flu: when China stopped live poultry sales in some cities during H7N9 bird flu outbreaks, the outbreaks stopped. So it is good news that China says it is phasing out live poultry sales at its wet markets, albeit not closing the markets completely, as they are also an important source of fresh fruit and vegetables and other perfectly safe forms of food.[10] Possibly wild animal sales, which China banned for food but not other purposes, could be cleaned up, too, if there is too much vested interest to stop them entirely. Better hygiene and animal welfare, and a stringently enforced ban on bats and endangered species, might raise costs enough to make the industry smaller but safer. All the same, if it still threatens biodiversity, it will still raise our risk of zoonosis.

Meanwhile, people in Africa who depend on fruit bats for protein pose a dilemma that could perhaps be addressed with respect and research, and agricultural and economic development. And there are probably other dangerous species lurking in wet markets. Yes, such markets have a long tradition, but those years of history didn't take place alongside megacities or our hyperconnected world, which magnify the risks they pose of swapping pathogens among species, including humans. We can find ways to provide needed goods safely.

Speaking of flu, in Chapter 5 we looked at the one virus we know is going to stage a pandemic, how it does this, and how the swine flu pandemic in 2009 precipitated an attack on the WHO that made it harder to react to Covid-19. Meanwhile, many countries struggled to respond to Covid-19 because their only pandemic planning was for flu, which requires a different kind of response. We still need those plans (and more), though, if only for bird flu, which is highly lethal and might be able to go pandemic in people while maintaining a death rate that would make Covid-19 look like the common cold. Of course, if another relatively mild pandemic flu is next to win the genetic lottery, we need to respond to that, too, without screaming that this whole flu thing is overhyped. It isn't.

WHO officials have suggested we might need a more nuanced response to different severities of pandemic flu, or other outbreaks judged to merit a WHO emergency declaration, to avoid the criticisms the response to the 2009 flu pandemic attracted. It's worth remembering, though, that however a virus may start out, we can't predict how it will evolve. And as Covid-19 demonstrated, even a virus that is less lethal than many can be unexpectedly dangerous. Some argue that a more lethal virus might, paradoxically, have done less damage, as there would have been less delay and hesitation in our efforts to contain it.

Meanwhile, strife over the risks of lab work to explore the pandemic potential of bird flu viruses suggests we should monitor high-containment labs more closely, transparently, and internationally. A good example of how *not* to do that occurred in April 2020, when US funding for research involving the bat coronavirus lab in Wuhan was summarily halted.

The lab's chief scientist, Shi Zhengli, has said the genetic sequence of the virus that causes Covid-19 does not match any they have sequenced.[11] Unsequenced virus from a bat sample, or an actual bat, could, in theory, have infected someone in the lab if there was a failure to apply the rules requiring stringent containment—and there are still unanswered questions about a bat collection and a huge waste clear-out just before the pandemic started, at the Hubei CDC lab in downtown Wuhan. But we know bat coronaviruses infect ordinary people living near bat colonies and could well infect people who collect, sell, and use bats or bat feces. That seems by far the bigger risk. Ultimately, if the aim is to be safer in future and not just to point fingers of blame, we don't really need to know if SARS-2 came from a bat in nature or via some medicinal or laboratory collection of bat material. The next virus from bats, or indeed other creatures, could come from any of those sources, so all must be managed better.

One thing we need is a transparent international system of inspection and accountability for these types of labs and open international decision-making about which research is, and is not, worth the risk, to ensure that important work to defend us from risky viruses is done, and done safely, and that labs are not randomly accused when disease emerges. We desperately need the labs and a lot of the science they do, more than ever, but when the risk is global, control and responsibility should be as well.

One thing we can say for sure: Covid-19 was not created in a lab. In an analysis in the prestigious journal *Nature Medicine* published in March 2020, scientists admitted that we simply would not have known enough to do it. We wouldn't even have guessed that the bit of protein Covid-19 uses to attach to human cells would work so well.[12] It does, though.

So, *lesson for the future number 5*: a flu pandemic is coming, and it is unlikely that it will take a century to do so—it could happen as early as this year. By now, it should surprise no one to hear that we aren't ready for a bad one. We've already done a lot of the homework on pandemic preparedness for flu, though, and those plans should now be revised in light of the hard lessons we are learning about pandemics from Covid-19—and actually be made ready to roll. At the same time, the WHO's Global Influenza Surveillance and Response System that monitors flu evolution should not only be maintained, but more generously funded and expanded to include other worrying virus families. An international collaborative effort among scientists who study the evolution of pathogens and livestock scientists—who are now rarely even in the same room—should aim to wean farm animals off some vaccines, antibiotics, and other management practices that foster dangerous pathogens—and resistance to this in agricultural industries should be shut down like any vested interest that puts people at risk. We need similar efforts to increase the transparency and safety of high-risk virus research. And we desperately need ways to make seasonal or pandemic flu vaccines for people much, much more quickly—and, if possible, we need a universal flu vaccine as well. Yes, that's right, both kinds of vaccine, belt and suspenders. Flu's a nasty customer: it deserves it.

After our trip through the long record of largely disregarded warnings that led to this moment, we looked in Chapter 6 at what we should be doing to prevent the next pandemic—whatever it is—or to respond and contain it fast if one starts. We need decent pandemic plans. We need stockpiles of response equipment. We need worldwide surveillance for emerging disease, as much as possible by local experts who understand their own situation but have a global network of colleagues and resources at their backs. We need a lot more basic work on diagnostic, vaccine, and drug technologies, and we need to deploy the capabilities we have so we are ready to use them, fast, and everywhere. It sounds expensive, but as we are learning, it will almost certainly cost way, way less than the next pandemic—if we can still muster the organization, and the cash, after this one to take the precautions we know we need.

*Lesson for the future number 6*: we need to hold governments accountable for their promises, during the crisis, to do all this. Actually, this lesson is one we can act on now. The G20 group of the world's biggest economies promised to take action on pandemics in late March 2020, to create "a universal, efficient, sustained funding and coordination platform to accelerate the development and delivery of vaccines, diagnostics and treatments."[13] They failed to say much about that when they had their next summit the following November, and as I write in mid-2021, I see no sign of it. Two-thirds of all the people in the world live in G20 countries—so more likely than not, dear reader, one of the governments that made that promise was yours. It needs to be held to account. Try to do that.

If we don't hold those responsible accountable now, we might see something like our trip to the dark side in Chapter 7. Few people realize that the rapidly increasing complexity of our global-ized society is increasing risk in ways that could be catastrophic. That is how the outbreak of a new respiratory virus in China rap-idly went pandemic and why painful economic domino effects just from trying to slow its spread have propagated globally. We looked at the possibility of a much worse pandemic, with a higher death rate, and discovered that the widespread belief that diseases that go pandemic always become milder is a myth. In fact, if we aren't careful, the wrong kind of vaccination can make a virus worse—although I should stress that vaccination in general is a Good Thing, mainly because it is always done very carefully. But to do that we need to understand what can sometimes go wrong.

I looked at pathogen evolution and complex systems together, not just because both are frightening, but because it's the two together that pose the real threat: a pathogen that evolves to become a more severe pandemic could precipitate cascading fail-ure in our complex global support systems. Especially if the low-income people who hold a lot of it together are further weakened by growing economic inequality—and if the shocking health pen-alties imposed on many racial minorities, who are also dispropor-tionately essential workers, are allowed to continue.

*Lesson for the future number 7*: we know pandemics are serious, but we are more vulnerable than many of us believe. After weath-ering the initial onslaught of Covid-19, we can't go back to normal. Normal is what led to this, and more of it means there will be more pandemics, and they could well be worse. We have to take the obvi-ous preventive measures of Chapter 6: stockpile PPE, build vaccine plants, do more disease surveillance, and plan. But the possibility that a big disease event will trigger epidemics of collapse throughout our global systems—food, water, security, financial, even nuclear—is the bigger problem we must try to fix, because that tightly linked complexity is why the risk, both of triggering pandemics and of their impacts, is increasing. If nothing else, an unflinching appraisal of the real risks might finally inspire action.

We must manage our global system with some understanding of how complexity works, taking advantage of the global shock caused by this pandemic to build looser connections, less efficiency, more redundancy, and resilience into global supply chains, economies, and governing structures, even if that is never the cheapest option. If a few connections collapse here and there, complex systems experts suggest it might be more opportunity than disaster: "creative destruction" might let new, more resilient patterns emerge, especially if we rebuild with that in mind.

We must grasp that a much worse pandemic can happen, and it could trigger nonlinear effects in our global system that could lead to the collapse of local systems or even global ones. Some of the world's smartest scientists say that is what we risk. Every disaster movie starts with someone ignoring a scientist.

So here we are. Are we back from the dark side now? Is there any good news?

Yes. Crisis can be an opportunity, even if that is badly translated Chinese. We desperately need to redesign the systems that failed to contain this pandemic if we are to, with luck, prevent or at least contain the next one.

If you take one thing away from all we have looked at in this book so far, it should be that people have been predicting this pandemic for decades, yet we were not prepared. Covid-19 was an unnecessary catastrophe: we knew enough to keep people away from bat viruses, to develop drugs and vaccines for coronaviruses, to plan better pandemic response, and to set up transparent, truly global surveillance networks for outbreaks of potentially pandemic disease. Such surveillance systems would mean that if an outbreak occurs despite prevention efforts, everyone would find out fast, whereupon pandemic plans would allow rapid, aggressive action to limit its spread.

But as Covid-19 emerged, one bureaucracy delayed the warning—and there was no international agency that could go in and verify what was happening on the ground, immediately, on behalf of everyone else. Then we didn't have the global public health infrastructure to ensure every country's response was adequate, even though inadequate response in any country could mean increased infection in others: it's no accident more dangerous variants of Covid first emerged in countries with poor pandemic response, hence more cases.[14] We didn't have crisis-management systems that could try to counteract local or national governments' denial and delay—even though that affected everyone.

The WHO did for Covid-19 a bit more than what in 2013 I predicted it would do for H7N9: it issued advice, held daily briefings, organized R&D, got PPE and test kits to poorer countries, and especially organized equitable global vaccine distribution. Yet most of the response was up to countries and too many were mired in hesitation and inaction.[15] "The virus is faster than our bureaucracy," admitted Italian authorities, as Covid overwhelmed their hospitals and what was needed was more like mobilizing for war.[16] Yet, fully a year later another government's inaction allowed Covid to devastatingly overwhelm health care in India.

So how do we fix that? Obviously, there must now be major investment in the scientific preparations we should have made for this pandemic. Jeremy Farrar, head of the Wellcome Trust in England, says countries need to invest in public health and in the clinical, social, and basic science of infectious disease. That includes many countries' long-atrophied capabilities to do basic epidemiological controls: isolation, quarantine, and contact tracing. Making our response and alert systems truly global might be less obvious. As I argued, we need a high-level, authoritative system bringing countries together to collaborate on disease so that no one conceals important details about worrying outbreaks and everyone works together. Easier said than done, perhaps, but where can we start?

Many criticize the WHO, although frankly I think it just makes an easy target. There were certainly things it might have done better: I think it could have called Covid-19 a public health emergency, and then admitted it was a pandemic, earlier, communicating the real urgency of the situation rather than holding back for fear of scaring people—or perhaps offending some governments. But the WHO has few choices in this regard: it can do very little independently of its member states. Yet it remains the world's only global health agency, and it must be part of building a better system.

First, let's look at why we have to organize globally at all. Globalization has become a bad word in some circles. Indeed, in Chapter 7, we looked at how a lot of our vulnerability in a pandemic is due to our tightly interlinked global systems. But the bad part of that is not the "global" part. It is the part where it is all so tightly and efficiently linked. That optimizes profits, but it also creates a rigid network that transmits shocks. In this pandemic, closed clothing stores in Europe created unemployment and hunger in Bangladesh, while factory shutdowns in China threatened the availability of essential drugs in the United States. Some experts think the tightly coupled, fragile global financial system, barely ten years removed from its last worldwide crisis, has come close to meltdown, too.[17]

But is the answer to make unemployment in Bangladesh permanent by bringing those clothing factories "home" or to shut down the amicable global trade links that have helped foster the longest stretch of relative peace the world has ever known?[18] If Covid-19 teaches us anything, it is that we really are all in this together.

Some people in the anti-globalist (or just plain nationalist) camp strongly believe we should not be organizing ourselves on a planetary scale at all. Yet given that virtually all our economic and cultural activity is now on that scale, or at least has global ramifications, it is hard to argue we should not also be managing our affairs on that level. Just having eight billion people filling virtually every available niche on this planet makes us global whether we like it or not. We can no longer run our affairs as if countries were each totally independent, while even a small fraction of us might do things that severely affect everyone: besides disease, there are greenhouse gases, ozone-depleting chemicals, overfishing, financial instability, pollution, deforestation, cybersecurity, nuclear weapons—the list goes on. To even try to get ahead of the cascading failures that can result—like pandemics—we have no choice but to organize on a global scale as well. If the pandemic doesn't teach us that—well, global warming will, but possibly not until it's too late for lessons.

"We've created a tightly knit socio-ecological system that reaches into all corners of the planet," says complexity expert Thomas Homer-Dixon. "If we're to grasp the nature of today's emerging global dangers and adequately mobilize ourselves to do something about them, 'we' needs to come to mean, to a lot of people most of the time, the entire human species."

So how do we do that, at least as it applies to preventing or responding to pandemics? Certainly, part of the answer must be strengthening the WHO so it can do the monitoring and coordinating job we are already asking of it, acting as a kind of global civil service for health, if not a real political authority. The nation-states that hold most of the power in the world seem unlikely to permit an international agency to wield comparable power of its own.

But we could at least allow it enough power and resources to play the supporting role effectively. As it stands, the WHO, the world's only organization charged with stopping pandemics and pursuing all other aspects of health that have an international dimension, had a budget of $2.4 billion per year for 2020 and 2021, virtually no real increase from the previous four years—and that was after a 20 percent cut in 2011 due to the financial crisis, with emergency and epidemic funding cut still further.

Yet after 2014, the WHO acquired an emergency response capability, expanded its work on antibiotic resistance and the health threats of climate change, almost completed polio eradication, and led the world's response to the Covid-19 pandemic. It was working on a shoestring as it was. With no increase in funding in that time, it is stretched pretty thin.

The worst part of this is that to get anything done, the WHO cannot afford to antagonize its more powerful member states that can simply refuse to cooperate. On January 23, 2020, it called an emergency committee to discuss the pneumonia China had just admitted was contagious. Yet even though there were known cases in several countries besides China, the committee was split between those who wanted to declare it a Public Health Emergency of International Concern, as the IHR demands, and those who didn't—led by China, which clearly opposed it. WHO didn't declare a PHEIC until a week later. "As in other recent outbreaks," the US Council on Foreign Relations, a think tank, later complained, "WHO prioritized solidarity in its international crisis response, proving hesitant to declare a PHEIC over China's objections."[19]

Then in April, former president Donald Trump protested the WHO's apparent deference to China by storming out of the agency, taking with him a large chunk of its funding—and, ironically, perfectly illustrating why the WHO had to move so carefully with China, for fear it might have done the same. The WHO, the CFR concluded, can and should lead global pandemic response, "but it is beleaguered, overstretched and underfunded," cannot ensure countries abide by the IHR, and "lacks the geopolitical heft to address the broader diplomatic, economic and security implications of pandemics."

Larry Gostin, an American expert in public health law, called Trump's exit an effort to deflect blame for the slow US response to the pandemic, even though the WHO had been screaming for weeks that countries needed to do more. Gostin charged that the WHO's budget is a third that of the CDC in the United States, and the CDC doesn't have to respond to health emergencies across the planet. The WHO also runs a plethora of programs to strengthen health systems in poor countries, which, we should all now realize, benefits all of us.

Worse, it gets to spend only a third of its budget as it likes; the rest is earmarked by member states for pet projects. Its emergency fund is run on voluntary contributions, and after using most of that to contain an Ebola outbreak in the Democratic Republic of the Congo between 2018 and 2020, it had a risible $9 million left to help poor countries respond when Covid-19 hit. It took countries weeks to respond to its emergency appeal for pandemic funds. Gostin says that to act in the interests of the world, the WHO needs a doubling of regular funding, and that needs to be less subject to the partisan interests of the richest member states.

Mostly, though, Gostin said in early 2020, "we need to finally recognize that this novel coronavirus is the common enemy and unite as a global force to overcome it," and any such threats that follow it.[20] In May 2020, UN Secretary-General António Guterres echoed this, saying the virus spread out of control because "the world was not able to come together and to face Covid-19 in an articulated, coordinated way."[21] The astronomical costs of the pandemic might finally ram home the idea that events with potentially catastrophic global effects should be a shared, global responsibility, and not be subject to the interests—or just the local medical or bureaucratic limitations—of any one country.

So, how about that thing we suggested earlier: a high-level, authoritative system bringing countries together to collaborate on disease so that no one conceals important details about worrying outbreaks and everyone works together? Right now, most power rests with sovereign nation-states, especially the twenty or so richest, most powerful ones. We saw how national sovereignty finally yielded to global health security, at least in a limited way, with SARS. But the WHO is still very much the creature of its 194 member states. When a country's interests do not coincide with those of the world at large, the WHO represents the interests of the world, but the country can often win.

The most obvious example is China's insistence in early January 2020 that Covid-19 was not contagious. But you don't have to be big and powerful: in 2014, the WHO's response to the Ebola epidemic in West Africa was delayed when the government of Guinea was reluctant to report true case numbers for fear of discouraging foreign investment.[22]

I would like to suggest two kinds of solutions. Neither one involves replacing the WHO—as I said earlier, it is the only game in town. In fact we need a stronger WHO that can act in the interests of the world despite occasional conflicts between the world's interests and the claims or abilities of nation-states. How do we do that? One way is to start from the recognition that we mentioned earlier: nation-states seem unlikely to give an international agency the power it needs to overrule nation-states. So if countries have all the power, they have to find some way to use it in the common interest. If the WHO cannot tell a big country what to do, other countries will have to.

A global government, in the top-down way that governments are usually understood, is not likely to work. Complexity scientist Yaneer Bar-Yam says that when social systems get too complex, old-fashioned hierarchies, with one guy (it's usually a guy) in charge, don't work anymore because one person can't get their head around everything. Hierarchies are already devolving into global networks as the real power structures in many areas of global concern, writes author and governance expert Anne-Marie Slaughter, especially things that can be managed by networks of experts like the development of shared international legal norms.[23]

So we need a network. We now have an annual meeting at which WHO member states tell the WHO what they want it to do. What if we also had a more constantly convened, high-level council of countries to deal with global health threats that would have the authority to demand that individual governments act, in close conjunction with the WHO, on big deals like suspicions of incipient pandemics, antibiotic resistance, and other problems that could have impacts far beyond one country? Where would we find a body like that?

The Council on Foreign Relations suggests a Global Health Security Coordination Committee of "like-minded states" to coordinate and "incentivize" investment in pandemic preparedness and to build an international surveillance network to exchange medical data and track emerging diseases.[24] But it isn't clear how such a voluntary association would get all countries to cooperate.

Maria Espona, an Argentinian expert on the Biological Weapons Convention, says that under the confidence-building measures of that treaty, member states—and as with the chemical weapons treaty, that is pretty much everyone—are supposed to submit regular reports of unusual disease outbreaks. As always, governments are unenthusiastic about airing their disease-related linen in public, so not many do, and the few reports submitted are spotty at best.

She suggests an automated reporting system in every country, where doctors, vets, and other disease experts report unusual events, and these go straight to a national automated system—no one has to fill out any forms, and the system takes care of formatting and categorizing.

Espona suggests putting this under the direct authority of the UN Security Council. The WHO, the UN Food and Agriculture Organization, the World Organisation for Animal Health, and civil society groups like the EcoHealth Alliance or Doctors Without Borders could form the essential network of technical support.[25] But the authority would come from nation-states working together— with enough clout to discourage secrets. Of course, the local inputs to such a system could be turned off by a government wanting to conceal an outbreak, as China did with their very similar system in Wuhan. But with the highest possible eyes on the situation—and perhaps artificial intelligence in the system to spot unconvincing reports—countries would have to either stop concealing outbreaks, or be answerable in some serious way if an unreported one does cause trouble. With intensified international monitoring of pathogens and open sharing of data, we might quickly learn which genetic variants occur where, making it harder to dodge responsibility for an outbreak. The UN Security Council is rightly criticized for often being paralyzed by dueling vetoes. But perhaps we could reimagine veto power as it concerns disease—and maybe make it unthinkable to be seen vetoing global response to what might become a pandemic. After all, even at the height of the Cold War, the United States and Soviet Union collaborated on smallpox eradication.

There is a second kind of solution we might consider that could fix some of the problems of the first: treaties. The 2005 revision of the International Health Regulations was an enormous improvement on what came before, which required member states to report on outbreaks of, by that time, only three diseases—which, true to form, they frequently did not do. In 2005, the new treaty charged countries with developing the capacity to detect, report, and respond to any serious, unexpected, or potentially international health threat, and to alert the WHO. And it let the WHO, not the country, decide if the outbreak was a Public Health Emergency of International Concern and recommend response measures, a hitherto unheard-of imposition on a country's sovereignty.

Otherwise, it showed just how firmly in control the nation-states intended to remain. The treaty gave the WHO the power to ask a country about a disease outbreak it had heard about from sources other than the government concerned, such as ProMED or even other governments. Before that, it could ask only about outbreaks that government had already admitted. The revision also allowed the WHO to talk about an outbreak publicly without the permission of the country involved—if the outbreak was already public knowledge or it was spreading outside the original country. It can also declare an emergency without the approval of the government involved. Those might seem like remarkably innocuous concessions, but it took until 2005, after SARS had almost spun out of control, to get that much. And even that took hard negotiating.

But crucially, the WHO still can't investigate an outbreak directly unless the country invites it. "Verification" under the IHR consists of asking the government involved for information. It could not go to China to investigate China's claim that Covid-19 did not spread

287

person to person before Beijing admitted it did on January 20, 2020, even though an earlier revelation of this might have slowed or limited the pandemic.

It's not as if it didn't have an inkling of what was going on before then: evidence has emerged that the WHO was quietly warning countries that the new pneumonia might well spread person to person as early as January 10.[26] "I was concerned that there had been no report of further cases or any information about transmission, particularly information about possible human-to-human transmission, between the announcement of the outbreak by Wuhan authorities on December 31st and January 17th," says John Mackenzie of Curtin University in Perth, Australia, then a member of the WHO's emergency committee on Covid-19. Yet WHO could not say so officially until Beijing did, and could not visit to establish what was happening until China invited it in February.

Add to that the fact that countries haven't done even what the treaty does demand. The IHR is legally binding, and it requires countries to improve their surveillance and public health capability and to assess their own ability to detect and respond to outbreaks, yet many have not done this adequately. "The global pandemic alert system is not fit for purpose," the WHO's independent investigative panel tersely concluded in January 2021, saying it was "struck that the power of WHO to validate reports of disease outbreaks for their pandemic potential . . . is gravely limited."[27] The US Council on Foreign Relations suggested an IHR review conference, the kind of meeting that could open the treaty for revision, to discuss how to improve compliance with its reporting requirements.[28] But there are problems with this: opening up the IHR would risk losing some of its existing requirements, agreement could take years of wrangling, and then could not come into force for two years.

Some governments think a whole new treaty might be more to the point. In March 2021, the WHO and twenty-four heads of governments, including Britain, Indonesia, South Africa, and much of the EU, called for a "new international treaty for pandemic preparedness and response" to improve alert systems, data sharing, research, and the global production and distribution of vaccines, medicines, diagnostics, and PPE, making immunization "a global public good." At the WHO's annual assembly in May 2021, more than thirty countries, this time including the US and the entire EU, backed a meeting in November to discuss the treaty. These measures, as we have seen throughout this book, are exactly what we need. But of course, they warned, it would "require a sustained political, financial and societal commitment over many years."[29]

It is not clear that we have many years to build the global transparency and capabilities to handle the next pandemic better than we handled Covid. The fastest way forward, experts suggested in May 2021, might be an treaty in addition to the IHR—which cover all health events—that applies only to potential pandemics. It could cover independent verification of outbreaks, monitoring and enforcement of surveillance, rules for travel restriction, sharing pathogens and vaccines, and managing zoonotic risk.[30] There are approaches to at least some of those things that might be worked out relatively quickly, as they have precedents in other treaties. All we would need would be the political will to follow through.

The problem with all international treaties is that there is no way to enforce them: they transcend individual countries, but countries in theory and often in practice hold all the world's political power, so no one is big enough to do the enforcement. But that is the wrong way to think about treaties. Enforcement is not how treaties get things done: verification is. There are already treaties that are effectively enforceable because countries have set aside one tightly circumscribed area of national sovereignty in the name of global security. The world doesn't have to take the country's word that it is following the treaty's rules: international inspectors can come in and verify whether it is—or isn't—which amounts to a suspension of the country's sovereignty in that area. The main ones govern nuclear material and chemical weapons.

Members of the Nuclear Non-Proliferation Treaty must declare any uranium or plutonium that can be used for nuclear weapons, prove they haven't diverted or enriched enough to pose the threat of weapons production, and submit those declarations to verification inspections by the International Atomic Energy Agency, or IAEA. The arrangement works: the IAEA caught Iran cheating, twice, and imposed an inspections regime that was keeping it from stockpiling potentially weapons-grade uranium—until Donald Trump torpedoed the agreement in 2018. The five official nuclear powers have not yet given up their weapons as promised in the treaty, and four countries have acquired them despite it, but, weapons experts tell me, the world is nowhere near as awash in nuclear material and weapons as was looking likely when the treaty came into force in 1970. It has also made declaring, for international inspection, activities that could otherwise endanger everyone a routine part of global security.

The 1997 Chemical Weapons Convention (CWC) bans the production or stockpiling of certain weapons, such as nerve gas, and prohibits using any chemical as a weapon. Member states—everyone but Israel, Egypt, North Korea, and South Sudan—declare any facilities they have that could make these things, and inspectors from the treaty's Organisation for the Prohibition of Chemical Weapons (OPCW) verify they aren't, and also check ordinary chemical plants. There are holes in the verification regime, as new weapons are developed that don't figure on its monitoring list (Novichok, used in alleged Russian assassination attempts in 2018 and 2020, is an example), but it has generally worked, notably with the chemical disarming of Syria in 2013—although Syria resumed use of such weapons largely due to a paralyzed UN Security Council.[31] A similar treaty banning biological weapons was also supposed to be equipped with a verification protocol, mandating inspections of bio labs. Although the convention still stands, it is largely toothless without that protocol, which was torpedoed by the United States in 2001.

The CWC's real innovation is that someone can charge a member state with not declaring a chemical weapon, or using one illegally, and ask for a surprise "challenge" inspection. Treaty countries have all agreed to "anytime, anywhere" inspections with no right of refusal, except the United States, which passed a law allowing it to refuse.[32,33] No one has ever demanded a challenge inspection, although the OPCW's destruction of Syrian chemical weapons in 2013–2014 was one in all but name. In another kind of check on bad behavior, the 1987 Montreal Protocol to the treaty banning chemicals that destroy the earth's protective ozone layer allows member states to slap trade sanctions on countries that break it. It never has, but at one time at least we all agreed such a threat was appropriate.

Although these treaties have had rather spotted records, they have if nothing else established what weapons experts call a "norm" against these weapons and chemicals: we have all agreed that we're not supposed to have them. "It somehow has to become an international norm that you don't just let infectious diseases fester" without reporting them and doing something about them, says Amesh Adalja at Johns Hopkins. We already have a treaty requiring countries to declare any worrying outbreaks of disease, the International Health Regulations. But as we saw in Chapter 1, although China fulfilled its obligations by reporting the pneumonia outbreak in Wuhan, it delayed the release of the virus's sequence and claimed it was not contagious, in an apparent effort to limit domestic panic and international attention. As we also saw in Chapter 1, the pandemic might not have been as bad if China had reported everything earlier.

In Geneva in May, WHO director Tedros said any new pandemic treaty must "make countries more accountable to one another," because "the safety of the world's people cannot rely solely on the goodwill of governments." The question raised by Covid is, How do we do that?

The answer could be real verification. On the model of weapons treaties, the IHR could require countries to make broad declarations about infectious diseases on their territory, regularly if things are normal, as a matter of urgency if there is a worrying outbreak. Then inspections could enable verification of what a country has said about its outbreaks, including that there are none. To believe a country's declaration that it has no worrying disease, we would need to know its local surveillance systems were able to spot one if there were.

That means verifying a country's surveillance systems in peacetime. There is already a similar type of verification in the program for polio eradication, in which the WHO participates: if a country says it hasn't found any cases of paralysis that turned out to be polio, it has to have found the number of cases of paralysis that weren't polio that you would normally expect in its population, so we know it was looking hard enough.[34] This kind of system could be generalized, in various ways, to all infectious disease, and verification inspections would uncover where local surveillance was still falling short and help bring it up to standard. Such a regular inspection regime might finally enable poor countries, with help from rich ones, to develop disease surveillance systems we could all rely on. More focus on health might also improve it in some countries—a bonus for everyone as healthier societies tend to be more prosperous and stable.[35]

We even have a ready-made verification agency: the WHO. Verifying outbreaks would have a completely different dynamic from a weapons treaty. After all, a country with banned weapons presumably acquired them on purpose and plans to use them, if only as a threat. A country can harbor a disease just through bad luck or difficult geography. And it usually isn't intending to launch it at an enemy: the virus will arrive in other countries anyway, on the next passenger flight. Inspectors in these circumstances are friends, not adversaries. Refusal of inspections, on the other hand, might raise suspicions of a cover-up or even a lab escape or bioweapons work.

In 2004, after China admitted (after some prompting) that it had H5N1 bird flu all over its territory, I wrote for *New Scientist* that we should "start controlling viruses the way we control nuclear weapons or ozone-depleting chemicals." The stakes, if anything, were even higher: were this flu to go pandemic, I wrote, "the economic cost, political toll and loss of human life would be colossal. Bring on the Pathogens Treaty."[36]

Years later, I am even more convinced that some agreement like this must be the answer—and perhaps the wreckage of Covid-19 will make it politically feasible. Pandemic disease, as Covid-19 has abundantly demonstrated, is more devastating to more countries for a longer time than any chemical weapon could ever be, yet countries have agreed to inspections "anytime, anywhere" to show they don't have chemical weapons and to make sure their neighbors don't. A pandemic actually can start anytime, anywhere—surely inspections that can keep pace with that and are accountable to everyone, not just the country involved, are the only defense any country can really trust.

I ran the idea past a few weapons experts. They think the world has grown tired of treaties, that "multilateral" cooperation among countries is no longer in vogue. Well, we're having a multilateral pandemic. Vogues change. Abstract concepts of national sovereignty might motivate treaty negotiators in conference rooms in Geneva, but in practice, with an unknown disease threatening, no country wants to look unreasonable to their partners in a global market over something that could threaten everyone. What we need is an authoritative way to shine a spotlight. If some country had an outbreak, and the WHO asked to come investigate, and that country said no, how would that look?

Such an arrangement might also help us head off the inevitable blame games governments play regarding disease. It would give a country a chance to defend itself against charges that a pathogen of concern escaped from a lab or, from the point of view of the country's neighbors, a chance to verify whether or not it did. And the prospect of inspections might make such escapes less likely. In Chapter 6, we looked at risky pathogen research that might be made safer if it came under transparent international supervision. That could be part of this treaty—or perhaps we should write a new verification protocol for the biological weapons treaty.

Besides, treaties are not all about carrying a big stick. They are also loaded with carrots, like the pledge in the IHR for rich countries to help poor ones monitor disease, and pledges in the nuclear, biological, and chemical weapons treaties for the rich to help poorer countries use those technologies peacefully. In these weapons treaties, treaty members conduct confidence-building exercises, where experts from other countries visit your installations and you visit theirs. And unlike governments, experts communicate easily, so cultures of secrecy are less likely: the virologist in Saudi Arabia who first encountered MERS solved the problem quickly by turning to a Dutch virologist, who in turn supplied the virus to qualified labs elsewhere for research or diagnostics. Everyone benefitted. Some Saudi authorities objected—but if there were an international arrangement that fosters such exchanges, of pathogens as well as data, makes them safe and transparent, fairly assigns any intellectual property rights, and makes any government objections look dangerously old-fashioned, the reasons for those complaints should not exist, and we all could get a much firmer grip on global disease.

The need for international solidarity on disease was growing even before Covid: Peter Piot, formerly the head of UNAIDS, told me that by 2030 Zambia will spend 3 percent of its GDP just to deal with HIV, and it will need help.[37] Moral considerations aside, why should rich countries care about HIV in Zambia? For the same reason they care about whether another Covid-19 might be brewing somewhere: less disease means less poverty, and that in turn means less risk of emerging disease, as people become more prosperous and stop having to take health risks just to live.

In fact, we can't talk about solving the problem of pandemics without talking about global inequality generally. Covid-19 came from China, which is not a poor country, but the same cannot be said of areas prone to other worrying viruses, from Ebola to Nipah to something unknown now because it lives in a country with no virology or disease surveillance going on. If the destruction of nature is what causes our increasing exposure to novel germs in animals, poverty and inequality are what cause that destruction, whether it's done by poor people trying to get food or money to live, or by rich people profiting from dangerous industries in countries too poor (or corrupt) to stop them. And we're talking greenhouse emissions here, not just chopping down bat forests.

In March 2020, at the G20 meeting that pledged pandemic preparedness measures, UN Secretary-General António Guterres insightfully said that to prevent any future pandemics, "we must work together now to set the stage for a recovery that builds a more sustainable, inclusive and equitable economy."[38] British columnist Tim Walker of *The New European* newspaper was hoping for much the same when he tweeted, as the pandemic took hold in early 2020, "When this is over, we may have got used to better air, seen the point to international co-operation, that people don't have to sleep on the streets. . . . We might twig there's more to life than nationalism and the economy. It could be a new beginning."[39]

Jonathan Weigel and his colleagues at the London School of Economics call for a global solidarity fund for pandemic response and recovery in poor countries. "The developed world cannot heal if the rest of the world is in critical care," they write. "Renewed commitment to multilateralism and global solidarity is the safest path forward—for all of us."[40]

Be it a global fund or a treaty pledging to monitor disease— and verify it—the imperative is clear, both for defeating Covid-19 and ensuring a human future less threatened by disease. And perhaps, given the cascading failures severe disease might trigger in our complex support systems, ensuring a human future at all. We really are all in this together, and we'd better start acting like it. And ideally, make it legally binding.

Unfortunately, thinking like that, involving international cooperation, may actually become more difficult as a result of this pandemic. In early 2021, Scott Gottlieb, former head of the US Food and Drug Administration, wrote that US national security agencies must get involved in disease surveillance, arguing that a better intelligence presence in China might have detected Covid in Wuhan in December 2019.[41] It is not clear how spies would have seen it when the China CDC had not, or what they could have done if they had.

Regardless, now that national security agencies have seen the security threat disease can pose, it seems likely that they could soon start recruiting public health graduates to their shiny new disease departments. In May 2021, US President Biden told US security agencies to look into the origin of SARS-2. Yet treating infectious disease as a national security issue and the domain of intelligence agencies could make the open sharing of surveillance, pathogens, and remedies more difficult. Making countries more secretive about disease is the exact opposite of what we need.

And if geopolitics isn't problem enough, there is psychology. Besides our biological immune system, psychologists have found evidence for a "behavioral immune system"—a tendency to avoid people who may be carrying disease. Besides avoiding obviously ill people, psychologists think humans universally tend to conform to their own in-groups and avoid people who are different, because originally, we were trying to avoid infection.[42]

This was actually a risk during our early evolution. When we were wandering hunter-gatherers and encountered another wandering tribe, the strangers might well have encountered different diseases and be carrying germs to which they had acquired resistance, but we had not. This was especially true because some aspects of disease resistance are genetic, and back then we would have shared fewer genes with another wandering tribe than we do now with fellow city dwellers. The disease risks that long-separate populations may pose to each other were confirmed with a vengeance when most of the native people of the Americas died of European diseases after Columbus arrived. Europeans got syphilis back.

There is evidence the behavioral immune system underlies the regrettably universal human tendencies toward tribalism and xenophobia—and this affects some people more than others. People with stronger disgust responses to descriptions and pictures of things that might pose a disease risk, like dead cats or rotten food, tend to be more xenophobic and politically conservative, as are people from places with more pathogens, now or historically.[43,44,45] Former President Trump, who expressed aversion to nonwhite foreigners, also shunned fallen soldiers and the disabled, refused to admit his own illnesses, was reported to have an intense fear of germs and death, and insisted Covid doesn't kill "healthy" people—still a false claim of Covid denialists. This fits a pattern linking dislike of foreigners to an above-average fear of death and disease.

Researchers have also focused on "authoritarian personality," a set of personality characteristics, partly determined by genes, which include a desire for order, obedience, conformity, and cohesion within the in-group with which the person identifies. Having those characteristics made it more likely that a person would vote for Donald Trump in the United States, or Brexit in Britain, in 2016, significantly more than any other variable measured.[46] It, too, is linked to an above-average fear of disease.

Cambridge psychologist Leor Zmigrod has discovered that people who live in US states and cities with a higher prevalence of diseases you catch from humans—but not diseases you get from animals, like Lyme disease—are more likely to have authoritarian personalities and to have voted for Donald Trump. States with more pathogens also tended to have more laws that restrict minorities, such as LGBTQ people. No other variable, such as education or life expectancy, correlated as well.[47]

Other research has found that activating the behavioral immune system, either with reports of a real disease outbreak or with disgusting images, smells, or mentions of disease, makes people more authoritarian and rejecting of out-groups—for example, increasing opposition to gay marriage.[48] Canadian psychologist Mark Schaller, who coined the term "behavioral immune system," found that, in 2014, Americans were more likely to tell pollsters they would vote Republican after the appearance of Ebola cases in the United States, especially in places with a heightened interest in the disease as reflected by Google searches for "Ebola." The same went for conservative voting intentions among Canadians.[49]

This link fits with a history of sometimes violent xenophobia and hostility to strangers around epidemics. European cities slaughtered Jews and gypsies during the Black Death. In 1793, Philadelphia blamed a deadly yellow fever outbreak on traveling actors. White North Americans have blamed cholera on Irish immigrants, AIDS on Haitians, plague on Chinese immigrants—Honolulu burned its Chinatown—and SARS and Covid-19 on ethnic Chinese. As a presidential candidate, Donald Trump blamed Latin American immigrants for "tremendous infectious disease."[50] All the claims were groundless.

An association between foreigners and disease sadly persists. The United States responded to its first detected case of Covid by shutting its borders, even as it made little effort to track the virus already spreading within them. By mid-April 2020, every country in the world had imposed entry restrictions, the most extensive travel bans in history, which in lost tourism alone cost over 2 percent of global GDP. Yet in December 2020, epidemiologists calculated that, while entry restrictions often made sense early in 2020 in places where case numbers were still low and infected arrivals could make a difference, by September in many countries—except places with few or no cases—letting foreigners in made barely any difference, as they were not more likely to be infectious than locals.[51] Yet restrictions remained: the UK, with high and rising case numbers, quarantined arrivals from China, which by that point had virtually none.

"If Covid-19 elevates the allure of authoritarian ideologies, the effects could be long-lasting," Zmigrod told me. But this might be only one part of the pandemic's effect on our political attitudes. When you take what psychologists call terror management into account, it might cut two ways.

Sheldon Solomon, an originator of terror management theory, says we stave off fear of death when we face an obvious threat like a pandemic by either reducing our risk, say by distancing, or if our fear is too intense, by denying the threat: claiming Covid is a hoax or no worse than flu, rejecting masks, or claiming it kills only the weak. This helped polarize reaction to the pandemic, with more denial on the political right, where fear of disease and death is stronger.

But terror management has more subtle effects, involving our symbolic defeat of death by promoting things that outlast us—like our in-group. In one experiment, Americans who were asked for donations to charity outside a funeral home gave more than those asked down the street—but to American charities, not to foreigners.[52] Reminders of death can boost both solidarity and xenophobia.

But, Solomon and colleagues write, constant, subtle reminders of death—a pandemic, for example—can also make people double down on their own values, promoting their in-group. And values differ across ages and innate personalities.[53]

For example, even though George Floyd was, tragically, far from the first Black man in the United States to be murdered by a policeman, it happened as the pandemic was reminding millions of people of death. In the unprecedented Black Lives Matter demonstrations of 2020, Solomon suspects many people—especially the many demonstrators who were not Black—jumped at the chance "to feel they are doing something of value in their lives" and affirm their identity as opposing racism. Others responded with increased need for authoritarian leadership and hostility to out-groups, which may partly explain extreme police responses to the demonstrations, QAnon, and the invasion of the US Congress in January 2021, or why so many voted for Donald Trump in 2020 despite his mismanagement of the pandemic.

This polarization is troubling when the world needs greater collaboration, not less, to defeat the shared risk of disease. To short-circuit these unconscious defenses, Michael Matthews, a US military psychologist—combat elicits very similar reactions—suggests fostering things that limit stress during the pandemic, including stronger social relationships and renewed purpose. Solomon suggests consciously acknowledging your fear of death, which may also help to redefine your in-group. "Recognizing that the coronavirus poses the same existential threat for all of us helps underscore that humanity is a group we all belong to," he says, and humanity is the in-group we should be supporting to deal with a fear of death.[54] It may or may not be possible, but really taking our shared risk to heart may be the only way to achieve Homer-Dixon's idea of "we."

The least likely prospect for cooperation now, though, seems to be between the United States and China, which have repeatedly traded barbs over the virus. Yet the need is great. In February 2020,

Shi Zhengli, Kevin Olival, and twenty-one other emerging-disease researchers made a detailed case for the United States and China to work "synergistically" on research into pandemic threats. Only better understanding of disease ecology, they wrote, "can avert the increasing numbers of catastrophes in waiting."[55]

The two countries between them dominate livestock production and the global trade in wild mammals, two major pathogen sources. China is the world's largest maker and consumer of antibiotics, more than half used in animals, hence a major source of resistant bacteria. The world's two biggest economies have a moral responsibility as "major drivers of the ecological change responsible for the emergence of new disease," the scientists argued—and, they noted, they also happen to have the world's biggest combined infrastructure for infectious disease research. Yet increased collaboration between the two could become less likely if disease really does promote authoritarian and xenophobic tendencies.

Optimists, however, are hoping that the shared threat, anxiety, and hardships many of us are experiencing will outweigh atavistic fears of infection and breed social solidarity instead of hyperactive behavioral immune systems. American author Rebecca Solnit has documented that, in the wake of many disasters, survivors support each other with generosity, resourcefulness, and altruism.[56] I find myself repeating the catchphrase "we are all in this together" in descriptions of the pandemic, because events have overwhelmingly shown the truth of that, for better and worse.

The pandemic "could help catalyze an urgently needed tipping event in humanity's collective moral values, priorities and sense of self and community. It could remind us of our common fate on a small, crowded planet," hopes Homer-Dixon. "We won't address this challenge effectively if we retreat into our tribal identities. Covid-19 is a collective problem that requires global collective action—just like climate change."

Whether renewed xenophobia or recognition of our shared peril dominates the world's response to the pandemic may depend on how countries deal with one thing: the apparent likelihood that Covid-19 started in China. Both the United States and China have accused the other of originating the virus.[57] Some American companies have launched lawsuits against China for covering up details about the disease, and China has persistently speculated that the virus could have originated outside China and arrived in Wuhan, possibly, in frozen food imports.[58]

Not everyone sees this in adversarial terms, however. In April 2020, 101 top US scholars and former officials, including such senior figures as Madeleine Albright and Susan Rice, petitioned the US government to cooperate with China in fighting Covid-19. "China has much to answer for in its response to the coronavirus: its initial coverup, its continuing lack of transparency," they wrote. "Notwithstanding this, we the undersigned believe that the logic for cooperation is compelling."[59]

In May, Ursula von der Leyen, president of the European Commission, called for an international independent inquiry into the origin of the virus, not with a view to assigning blame, but so the world can work together to prevent it from happening again. "It's in our own interest, of every country, that we are better prepared the next time," she insisted, calling as well for a "transparent" early warning system. "The whole world has to contribute to that."[60] Her message: we need China to be part of this effort, and blame games will not help.

While it is true that China hid details of the virus from the world for a crucial few weeks, it is also true that China itself sustained enormous economic damage, which as in other countries was caused mostly by its efforts to stop the spread of the virus. That saved China's economy but also bought the world time. Nor was it the only country that was slow in recognizing and responding to Covid-19. Lots of countries, including the United States, made mistakes that hardly need enumerating here. Recognition of all that by all sides, including by China, might lead to the kind of common purpose that would prevent this from happening again.

Pathogens come from all over. The last flu pandemic started on an American-owned farm in Mexico, the biggest-ever Ebola epidemic began with the infection of a two-year-old child in one of the poorest countries in Africa. The HIV pandemic was seeded in an African society upended by European colonialism. The Zika virus started in Africa and then traveled via Asia, Micronesia, and Polynesia to Brazil, then wherever in the Americas it could find the right mosquitoes, which were themselves transported worldwide by numerous countries. Of the viruses that are still mere threats, Nipah started in Malaysia and the very similar Hendra in Australia. This is a planetary problem.

Jeremy Farrar seemed to be addressing those very concerns when he spoke at a virtual meeting organized in April 2020 by the US National Academy of Sciences. "Throughout history the world has faced great crises. In the aftermath there is always a choice," he said. "Do we apportion blame, exact reparations and become ever more polarized? Or do we come together, learn lessons, make changes and refashion a more collective, cohesive world?"[61]

Viruses don't care about human borders, identities, or ideologies—just human cells. The question now is, Do we care enough about defeating them to truly join forces?

# ACKNOWLEDGMENTS

This book was originally written as what the book trade calls a "crash" book, one written in a very short time at moments when a lot of people very much want to know about a certain issue. They are written by people who just happen to know a bit about the topic and are poised to go. That's the position I found myself in with Covid-19. I have spent most of my working hours since 1982 as a science journalist, mostly for *New Scientist*, a weekly scientific magazine based in London. Since the late 1980s, a large part of my beat has been infectious disease, the kind caused by germs rather than toxic chemicals or faulty genes—including viruses like Covid-19.

Obviously, writing a book in two months is not the way I always pictured my first outing as an author. Inevitably, there are going to be rougher edges than might have been the case if I had had the more traditional year or so to do this, even though I have, with this volume, been able to revise and expand what I wrote in 2020 from the perspective of early 2021. But whatever the limitations involved in a "crash" book, I couldn't pass up a chance to tell people what I have been hearing for years about emerging disease, just when people are really able to hear it. To colleagues who said I was crazy, you have a point. But now, months later, I still think we all need to take a step back and look at the big picture.

And that is what I have continued to do with this revised and expanded version, which I started updating seven months after the first edition came out. I hope I have smoothed some of the rough edges. I have certainly been able to include some of the mass of research that has emerged since last April.

But there are still many vitally important aspects of this pandemic I cannot write about properly in a project like this. Readers may be angry that I do not go into damning detail about this or that politician who screwed up their country's response and thereby caused deaths that might have been avoided. Yes, there have been those. Holding them to account will be massively important as the dust settles from this pandemic, if it does. I still can't do an analysis, certainly not one that would be appropriate for a book, of events that are still the stuff of news and changing by the hour. Many, many of my colleagues are doing that brilliantly elsewhere.

I also couldn't tell you what drugs and vaccines or economic and social remedies work, and why. I could start to tell you how some governments contained the virus while even more failed to, but I could not tell you yet how that plays out as the evolving virus comes back for repeated rounds. Those analyses will be happening, probably, for the rest of our lives and beyond.

What I could tell you is why we already knew this was going to happen and how the predictions panned out. I could tell you where we have got the wrong end of various sticks about pandemics. I could try to convey the big picture on the pandemic risks we run and, most importantly, what we should do, now, to try to stop this—and possibly worse—from happening again.

Obviously, I could not be writing this book if I hadn't been privileged to write about these subjects and allied ones for decades for a magazine as uniquely dedicated to the juncture between science and society as *New Scientist*. I could talk to the scientists and get that big picture because of the respect most of them hold for the magazine.

But getting the story out meant putting a lot of editors through some stressful moments. There were the features that took far too long as I dove down rabbit holes of unexpected twists in stories ranging from the impact of megadeath to the disappearance of Indian vultures. There were the countless news stories, week after week on harried deadline days, as news editors and subs fielded everything from last-minute shifts in the story as new facts emerged, to correctly spelling the names of obscure viruses and virologists, to occasionally standing up to a boss's cries of "Not another story about flu!"

There have been far too many editorial survivors of those deadline days to single any out by name, for fear of unfairly neglecting others. Apologies. But you know who you are. Thank you.

I would like to single out the editor who got me started as a news journalist, though. Back in the 1980s, I was the survivor of a bruising decade of grad schools and labs, and I had decided to write loftily about science for the masses instead.

Then Fred Pearce, the new news editor at *New Scientist*, heard I was based on the Continent and started sending me off to do news stories. I rapidly became hooked, and I'm grateful. Obviously, some of the information in this book was learned or unearthed in the course of investigations I did for *New Scientist* over the years, and I hereby totally acknowledge that. I note in the text when I am particularly indebted to a particular story.

I would especially like to thank the various scientists and allied experts who were patient enough to explain complicated bits of their life's work to me, at length and repeatedly, so I could write those stories—often as I, or they, were facing a screaming deadline. There have been only three I know of who refused to ever speak to me again after I wrote the story. And I don't regret those.

I would also like to thank the scientists who helped me with all the new science I had to digest to write this book, even though most of them were putting in long shifts and hazard duty on the front lines of Covid-19. All the mistakes that got through are, obviously, mine. I'm sure my scientist contacts will be letting me know about them.

I would like to express unlimited gratitude to my agent, Max Edwards, whose wild harebrained scheme this was—if they ever let us go to restaurants again, Max, I owe you serious lunch. Enormous thanks as well to the people at my publishers, Little Brown and Hachette Books, who shepherded this novice author through a locked-down book tour, and the publicists, editors, interviewers, lecturers, and fellow interviewees who made those Zoom discussions, phone-in shows, print articles, classroom Q&As, and even Reddit—the highlights of the long second half of 2020. Finally I would like to express my continuing respect and gratitude to my editor, Sam Raim, who took on the unlikely task of trying to get a book together and presentable in a ridiculously short time, all while working from home—then doing it again with this revised version. The mistakes that remain despite his desperate ministrations are mine. Except for all the Oxford commas: those are his.

Finally, like all authors, I need to thank my long-suffering family who had to deal with me disappearing into my office for weeks and muttering obsessively about disease when I emerged—and that was for the past several decades, although it admittedly got a lot worse during the frantic weeks I was writing this book, then writing it again. My family has been endlessly supportive, despite being locked down in various places as Covid-19 raged. That includes Smokey the cat, my office companion, who didn't quite make it to her nineteenth birthday as this revised book went through editing. Her politely lifted tail, poised sociably between keyboard and screen, was a constant presence as I wrote, and a source of much-needed cheer in Zoom sessions on Covid from C-SPAN to book festivals.

Thank you as well, of course, to my husband for constant cups of tea, watering the roses, and holding everything together. Thanks to both him and my daughters for playing the critical reading public with a few of these chapters, all while doing their jobs from lockdown—and in the case of my daughters, recovering from their own encounters with Covid-19. Now that you've got your sense of taste back, I still promise you carrot cakes till the cows come home.

# NOTES

## PREFACE TO THE REVISED AND EXPANDED EDITION

1 Larry Brilliant in conversation with Stewart Brand, "Sometimes Brilliant," *The Interval*, San Francisco, February 21, 2017, theinterval.org /salon-talks/02017/feb/21/sometimes-brilliant-conversation-stewart -brand.

2 Matt Apuzzo, Selam Gebrekidan, and David D. Kirkpatrick, "How the World Missed Covid-19's Silent Spread," *New York Times*, June 27, 2020, www.nytimes.com/2020/06/27/world/europe/coronavirus-spread -asymptomatic.html.

3 Lynne Peeples, "Face Masks: What the Data Say," *Nature* 586 (2020): 186–189, doi.org/10.1038/d41586-020-02801-8.

4 Robert Rosner, "Opinion: The Most Dangerous Situation Humanity Has Ever Faced," CNN, January 27, 2021, edition.cnn.com/2021/01/27/opi nions/doomsday-clock-dangerous-situation-brown-rosner/index.html.

5 Oxfam International, "Half a Billion People Could Be Pushed into Poverty by Coronavirus, Warns Oxfam," July 13, 2020, www.oxfam.org/en /press-releases/half-billion-people-could-be-pushed-poverty-corona virus-warns-oxfam.

6 World Bank, "COVID-19 to Add as Many as 150 Million Extreme Poor by 2021," October 7, 2020, www.worldbank.org/en/news/press-release /2020/10/07/covid-19-to-add-as-many-as-150-million-extreme-poor -by-2021.

7 Oxfam, "Mega-rich Recoup COVID-Losses in Record-Time yet Billions Will Live in Poverty for at Least a Decade," January 25, 2021, www

.oxfam.org/en/press-releases/mega-rich-recoup-covid-losses-record
-time-yet-billions-will-live-poverty-least.

8 Martin Wolf, "What the World Can Learn from the Covid-19 Pandemic," *Financial Times*, November 24, 2020, www.ft.com/content/7fb55
fa2-4aea-41a0-b4ea-ad1a51cb415f.

9 Phillip Alvelda, Thomas Ferguson, and John C. Mallery, "To Save
the Economy, Save People First," Institute for New Economic Thinking,
November 18, 2020, www.ineteconomics.org/perspectives/blog/to-save
-the-economy-save-people-first.

10 The Independent Panel for Pandemic Preparedness and Response,
"Second Report on Progress," January 2021, theindependentpanel.org
/wp-content/uploads/2021/01/Independent-Panel_Second-Report
-on-Progress_Final-15-Jan-2021.pdf.

11 Kamran Abbasi, "Covid-19: Social Murder, They Wrote—Elected,
Unaccountable, and Unrepentant," *BMJ* 327, no. 314 (February 2021), doi
.org/10.1136/bmj.n314.

12 David J. Scheffer, "Is It a Crime to Mishandle a Public Health
Response?," Council on Foreign Relations, April 22, 2020, www.cfr.org
/article/it-crime-mishandle-public-health-response.

**PREFACE**

1 Héctor Pifarré i Arolas et al., "Years of Life Lost to COVID-19 in 81
Countries," *Scientific Reports* 11, no. 1 (February 2021), doi.org/10.1038
/s41598-021-83040-3.

2 Vineet D. Menachery et al., "A SARS-Like Cluster of Circulating Bat
Coronaviruses Shows Potential for Human Emergence," *Nature Medicine*
21, no. 12 (November 2015): 1508–1513, doi.org/10.1038/nm.3985.

3 Debora MacKenzie, "Why We Are Sitting Ducks for China's Bird
Flu," *New Scientist*, May 1, 2013, www.newscientist.com/article/mg21829
150-200-why-we-are-sitting-ducks-for-chinas-bird-flu.

4 Institute of Medicine (US) Committee on Emerging Microbial Threats
to Health, *Emerging Infections: Microbial Threats to Health in the United
States*, ed. Joshua Lederberg, Robert E. Shope, and Stanley C. Oaks Jr.
(Washington, DC: National Academies Press, 1992), doi.org/10.17226/2008.

## CHAPTER 1: COULD WE HAVE STOPPED THIS WHOLE THING AT THE START?

1 ProMED-mail, "Undiagnosed pneumonia—China (HU): RFI," *ProMED-mail Archive 20191230.6864153*, December 30, 2019, www.promedmail.org. (Brackets are in the original text.)

2 ProMED-mail, "Undiagnosed viral pneumonia—China: (AH) medical staff, RFI," *ProMED-mail Archive 20130614.1773873*, June 14, 2013, www.promedmail.org.

3 Elisabeth Rosenthal with Lawrence K. Altman, "China raises tally of cases and deaths in mystery illness," *New York Times*, March 27, 2003, www.nytimes.com/2003/03/27/world/china-raises-tally-of-cases-and-deaths-in-mystery-illness.html.

4 WHO, "Timeline: WHO's COVID-19 response," www.who.int/emergencies/diseases/novel-coronavirus-2019/interactive-timeline.

5 Josephine Ma, "Coronavirus: China's first confirmed Covid-19 case traced back to November 17," *South China Morning Post*, March 13, 2020, www.scmp.com/news/china/society/article/3074991/coronavirus-chinas-first-confirmed-covid-19-case-traced-back.

6 ProMED-mail, "Undiagnosed pneumonia—China (HU) (02): updates, other country responses, RFI," *ProMED-mail Archive 20200103.6869668*, January 3, 2020, www.promedmail.org.

7 World Health Organization, "Pneumonia of unknown cause—China," January 5, 2020, www.who.int/csr/don/05-january-2020-pneumonia-of-unkown-cause-china/en.

8 ProMED-mail, "Undiagnosed pneumonia—China (HU) (05): novel coronavirus identified," *ProMED-mail Archive 20200108.6877694*, January 8, 2020, www.promedmail.org.

9 Jeremy Farrar, Twitter post, January 10, 2020, 9:50 a.m., twitter.com/JeremyFarrar/status/1215647022893670401.

10 "独家丨新冠病毒基因测序溯源：警报是何时拉响的" Caixin, February 26, 2020, web.archive.org/web/20200227094018/http:/china.caixin.com/2020-02-26/101520972.html.

11 Li-Li Ren et al., "Identification of a novel coronavirus causing severe pneumonia in human: A descriptive study," *Chinese Medical Journal* 133, no. 9 (May 5, 2020): 1015–1024, doi.org/10.1097/CM9.0000000000000722.

# Notes

12 Translation by Elisabeth Bik, "Dr. Ai Fen, 艾芬, the Wuhan Whistle," *Scientific Integrity Digest*, March 11, 2020, scienceintegritydigest.com/2020/03/11/dr-ai-fen-the-wuhan-whistle.

13 James Kynge, Sun Yu, and Tom Hancock, "Coronavirus: The cost of China's public health cover-up," *Financial Times*, February 6, 2020, www.ft.com/content/fa83463a-4737-11ea-aeb3-955839e06441.

14 Keisuke Kawazu, "Public backlash over China gov't accusations against docs who sounded coronavirus alarm," *The Mainichi*, January 31, 2020, mainichi.jp/english/articles/20200131/p2a/00m/0in/021000c.

15 Lotus Ruan, Jeffrey Knockel, and Masashi Crete-Nishihata, "Censored contagion: How information on the coronavirus is managed on Chinese social media," *The Citizen Lab* (University of Toronto), March 3, 2020, citizenlab.ca/2020/03/censored-contagion-how-information-on-the-coronavirus-is-managed-on-chinese-social-media.

16 Lily Kuo, "Coronavirus: Wuhan doctor speaks out against authorities," *Guardian*, March 11, 2020, www.theguardian.com/world/2020/mar/11/coronavirus-wuhan-doctor-ai-fen-speaks-out-against-authorities.

17 David Cyranoski, "Zhang Yongzhen: Genome sharer," from "Nature's 10: Ten people who helped shape science in 2020," *Nature*, December 14, 2020, www.nature.com/immersive/d41586-020-03435-6/index.html.

18 Wenji Tan et al., "Notes from the field: A novel coronavirus genome identified in a cluster of pneumonia cases—Wuhan, China 2019–2020," *China CDC Weekly* 2, no. 4 (2020): 61–62, doi.org/10.46234/ccdcw2020.017.

19 James Bandler et al., "Inside the fall of the CDC," *ProPublica*, October 15, 2020, www.propublica.org/article/inside-the-fall-of-the-cdc.

20 Lawrence Wright, "The plague year," *New Yorker*, December 28, 2020, www.newyorker.com/magazine/2021/01/04/the-plague-year.

21 Michael D. Shear, Sheri Fink, and Noah Weiland, "Inside Trump administration, debate raged over what to tell public," *New York Times*, March 9, 2020, www.nytimes.com/2020/03/07/us/politics/trump-coronavirus.html.

22 "Wuhan virus probably is spreading between people," *Radio Television Hong Kong*, January 4, 2020, news.rthk.hk/rthk/en/component/k2/1500994-20200104.htm.

# Notes

23 Kizzmekia S. Corbett et al., "SARS-CoV-2 mRNA vaccine design enabled by prototype pathogen preparedness," *Nature* 586 (August 2020): 567–571, doi.org/10.1038/s41586-020-2622-0.

24 WHO, "WHO statement on novel coronavirus in Thailand," January 13, 2020, www.who.int/news/item/13-01-2020-who-statement-on-novel-coronavirus-in-thailand.

25 "China didn't warn public of likely pandemic for 6 key days," Associated Press, April 15, 2020, apnews.com/article/68a9e1b91de4ffc166acd6012d82c2f9.

26 Natsuko Imai et al., "Report 1—Estimating the potential total number of novel Coronavirus (2019-nCoV) cases in Wuhan City, China," MRC Centre for Global Infectious Disease Analysis, January 17, 2020, www.imperial.ac.uk/mrc-global-infectious-disease-analysis/covid-19/report-1-case-estimates-of-covid-19.

27 Qun Li et al., "Early Transmission Dynamics in Wuhan, China, of Novel Coronavirus–Infected Pneumonia," *The New England Journal of Medicine* 382 (March 2020): 1199–1207, doi.org/10.1056/NEJMoa2001316.

28 *Sina*, news.sina.com.cn/s/2020-01-21/doc-iihnzhha3843904.shtml.

29 ProMED-mail, "Novel coronavirus (05): China (HU), Japan ex China," *ProMED-mail Archive 20200115.6891515*, January 15, 2020. Available at : www.promedmail.org.

30 *Sina*, news.sina.com.cn/s/2020-01-21/doc-iihnzhha3843904.shtml.

31 "China Didn't Warn," Associated Press.

32 ProMED-mail, "Novel coronavirus (07): China (HU), Thailand ex China, Japan ex China, WHO," *ProMED-mail Archive 20200117.6895647*, January 17, 2020. Available at: www.promed mail.org.

33 ProMED-mail, "Novel coronavirus (11): China (HU), South Korea ex China," *ProMED-mail Archive 20200120.6899007*, January 20, 2020. Available at: www.promedmail.org.

34 Caixin, www.caixin.com/2020-01-20/101506222.html.

35 Ma, "Coronavirus: China's first confirmed Covid-19 case traced back to November 17."

36 Jonathan Pekar et al., "Timing the SARS-CoV-2 Index Case," bioRxiv preprint, November 24, 2020, doi.org/10.1101/2020.11.20.392126.

## Notes

37 Josephine Ma and Zhuang Ping-hui, "5 million left Wuhan before lockdown, 1,000 new coronavirus cases expected in city," *South China Morning Post*, January 26, 2020, www.scmp.com/news/china/society/article/3047720/chinese-premier-li-keqiang-head-coronavirus-crisis-team-outbreak.

38 Huaiyu Tian et al., "An investigation of transmission control measures during the first 50 days of the COVID-19 epidemic in China," *Science*, March 31, 2020, doi.org/10.1126/science.abb6105.

39 Debora MacKenzie, "New coronavirus looks set to cause a pandemic—how do we control it?," January 29, 2020, www.newscientist.com/article/2231864-new-coronavirus-looks-set-to-cause-a-pandemic-how-do-we-control-it.

40 MacKenzie, "New coronavirus looks set to cause a pandemic—how do we control it?"

41 Chaolin Huang et al., "Clinical features of patients infected with 2019 novel coronavirus in Wuhan, China," *The Lancet* 395, no. 10223 (January 2020): 497–506, doi.org/10.1016/S0140-6736(20)30183-5.

42 Shengjie Lai et al., "Effect of non-pharmaceutical interventions for containing the COVID-19 outbreak: An observational and modelling study," medRxiv preprint, March 9, 2020, doi.org/10.1101/2020.03.03.20029843.

43 Enrique Rivero, "COVID-19 may have been in L.A. as early as last December, UCLA-led study suggests," *UCLA Newsroom*, September 10, 2020, newsroom.ucla.edu/releases/covid-may-have-been-in-la-as-early-as-december-2019.

44 The Independent Panel, "Second Report."

45 Steven Lee Myers, "China created a fail-safe system to track contagions. It failed," *New York Times*, March 29, 2020, www.nytimes.com/2020/03/29/world/asia/coronavirus-china.html.

46 "WHO-convened global study of origins of SARS-CoV-2: China Part (Joint WHO-China study: 14 January–10 February 2021)," March 30, 2021, www.who.int/publications/i/item/who-convened-global-study-of-origins-of-sars-cov-2-china-part.

47 Phil Hammond, Twitter post, January 24, 2020, 3:10 a.m., twitter.com/drphilhammond/status/1220619993408266241.

48 William A. Wagenaar and Sabato D. Sagaria, "Misperception of exponential growth," *Perception & Psychophysics* 18 (1975): 416–422, doi.org/10.3758/BF03204114.

# Notes

49 Joris Lammers, Jan Crusius, and Anne Gast, "Correcting misperceptions of exponential coronavirus growth increases support for social distancing," *PNAS* 117, no. 28 (July 2020): 16264–16266; doi.org/ 10.1073/pnas.2006048117.

50 Bob Woodward, *Rage* (New York: Simon & Schuster, 2020).

51 Helen Branswell, Twitter thread, February 11, 2021, twitter.com /helenbranswell/status/1359906760140730368.

52 Matt J. Keeling et al., "The efficacy of contact tracing for the containment of the 2019 novel coronavirus (COVID-19)," medRxiv preprint, February 17, 2020, doi.org/10.1101/2020.02.14.20023036.

53 Debora MacKenzie, "Why coronavirus superspreaders may mean we avoid a deadly pandemic," *New Scientist*, February 18, 2020, www.new scientist.com/article/2234388-why-coronavirus-superspreaders-may -mean-we-avoid-a-deadly-pandemic.

54 Joel Hellewell et al., "Feasibility of controlling COVID-19 outbreaks by isolation of cases and contacts," *The Lancet Global Health* 8 (February 2020): 488–496, doi.org/10.1016/S2214-109X(20)30074-7.

55 Pekar et al., "Timing the SARS-CoV-2 Index Case."

56 Kynge, Yu, and Hancock, "Coronavirus: The cost of China's public health cover-up."

57 Simiao Chen et al., "Fangcang shelter hospitals: A novel concept for responding to public health emergencies," *The Lancet* 395, no. 1032 (April 2020): 1305–1314, doi.org/10.1016/S0140-6736(20)30744-3.

58 "Report of the WHO-China Joint Mission on Coronavirus Disease 2019 (COVID-19)," February 2020, www.who.int/docs/default-source/corona viruse/who-china-joint-mission-on-covid-19-final-report.pdf.

59 Lee Hsien Loong, "PM Lee Hsien Loong on the 2019-nCoV situation in Singapore," Facebook, February 8, 2020, www.facebook.com/watch /?v=1284271178628870.

60 Aylin Woodward, "A new documentary shows how a top CDC official who warned Americans about the coronavirus promptly vanished from public view," *Business Insider France*, October 21, 2020, www .businessinsider.fr/us/cdc-official-warned-us-coronavirus-was-silenced -documentary-2020-10.

61 Lee Hsien Loong, "PM Lee Hsien Loong on the 2019-nCoV situation in Singapore."

62 Nana Addo Dankwa Akufo-Addo, "Full text of Akufo-Addo's fourth address on coronavirus pandemic," *JOY Online*, March 28, 2020, www.myjoy online.com/full-text-of-akufo-addos-fourth-address-on-coronavirus -pandemic.

63 Darren Lilleker, Ioana A. Coman, Miloš Gregor, Edoardo Novelli, eds., *Political Communication and COVID-19: Governance and Rhetoric in Times of Crisis* (Abingdon, UK: Routledge, 2021).

64 "COVID-19 deaths worldwide per million population as of March 11, 2021," Statista, www.statista.com/statistics/1104709/coronavirus-deaths -worldwide-per-million-inhabitants.

65 Benjamin J. Cowling et al., "Impact Assessment of Non-Pharmaceutical Interventions against Coronavirus Disease 2019 and Influenza in Hong Kong: an Observational Study," *The Lancet Public Health* 5, no. 5 (April 2020), doi .org/10.1016/s2468-2667(20)30090-6.

66 Andrea Crisanti and Antonio Cassone, "In one Italian town, we showed mass testing could eradicate the coronavirus," *Guardian*, March 20, 2020, www.theguardian.com/com mentisfree/2020/mar/20 /eradicated-coronavirus-mass-testing-covid-19-italy-vo.

67 "Eight Wuhan residents praised for 'whistle-blowing' virus outbreak," *Global Times*, January 29, 2020, www.globaltimes.cn/content/1177960.shtml.

68 Danilo Cereda et al., "The early phase of the COVID-19 outbreak in Lombardy, Italy," *arXiv* pre-print, March 20, 2020, arxiv.org/abs/2003.09320.

69 Giuseppina La Rosa et al., "SARS-CoV-2 has been circulating in northern Italy since December 2019: Evidence from environmental moni- toring," *The Science of the Total Environment* 750 (2021): doi.org/10.1016 /j.scitotenv.2020.141711.

## CHAPTER 2: WHAT ARE THESE EMERGING DISEASES, AND WHY ARE THEY EMERGING?

1 David S. Jones, "History in a crisis—lessons for Covid-19," *New England Journal of Medicine* 382, no. 18 (April 2020): 1681–1683, doi.org /10.1056/nejmp2004361.

2 Deborah MacKenzie, "Have we found the true cause of diabetes, stroke and Alzheimer's?," *New Scientist*, August 7, 2019, https://www

# Notes

.newscientist.com/article/mg24332420-900-have-we-found-the-true-cause-of-diabetes-stroke-and-alzheimers/.

3 "AJPH editorial: US readiness for COVID-19, other outbreaks hinges on investments to public health system," American Public Health Association, February 13, 2020, www.apha.org/news-and-media/news-releases/ajph-news-releases/2020/ajph-editorial.

4 Melinda Wenner Moyer, "A Wave of Resurgent Epidemics Has Hit the U.S.," *Scientific American*, May 1, 2018, www.scientificamerican.com/article/a-wave-of-resurgent-epidemics-has-hit-the-u-s.

5 Chris Thomas, "Hitting the poorest worst? How public health cuts have been experienced in England's most deprived communities," Institute for Public Policy Research, May 11, 2019, www.ippr.org/blog/public-health-cuts#anounce-of-prevention-is-worth-a-pound-of-cure.

6 Ab Osterhaus and Leslie Reperant, "Emerging and re-emerging viruses: Origins and drivers," European Society for Virology, April 11, 2016, www.eusv.eu/emerging-and-re-emerging-viruses-origins-and-drivers.

7 "Contagion: Historical Views of Diseases and Epidemics," Harvard Library, ocp.hul.harvard.edu/contagion/tuberculosis.html.

8 "Contagion: Historical Views of Diseases and Epidemics."

9 Rafael Lozano et al., "Global and regional mortality from 235 causes of death for 20 age groups in 1990 and 2010: A systematic analysis for the Global Burden of Disease Study 2010," *The Lancet* 380 (2012): 2095–2128, ipa-world.org/society-resources/code/images/95b1494-Lozano%20Mortality%20GBD2010.pdf.

10 Nuno R. Faria et al., "The early spread and epidemic ignition of HIV-1 in human populations," *Science* 346, no. 6205 (October 2014): 56–61, doi.org/10.1126/science.1256739.

11 Jacques Pépin, *The Origin of AIDS* (Cambridge: Cambridge University Press, 2011).

12 Institute of Medicine (US) Committee on Emerging Microbial Threats to Health, *Emerging Infections: Microbial Threats to Health in the United States*, ed. Joshua Lederberg, Robert E. Shope, and Stanley C. Oaks Jr. (Washington, DC: National Academies Press, 1992), doi.org/10.17226/2008.

13 Commission on a Global Health Risk Framework for the Future, National Academy of Medicine, Secretariat, *The Neglected Dimension of Global Security: A Framework to Counter Infectious Disease Crises* (Washington, DC: National Academies Press, 2016), doi.org/10.17226/21891.

14 Ariane Düx et al., "Measles Virus and Rinderpest Virus Divergence Dated to the Sixth Century BCE," *Science* 368, no. 6497 (June 2020): 1367–1370, doi.org/10.1126/science.aba9411.

15 Nathan D. Wolfe et al., "Origins of Major Human Infectious Diseases," *Nature* 447, no. 7142 (May 2007): 279–283, doi.org/10.1038/nature05775.

16 Debora MacKenzie, "Sick to death," *New Scientist*, August 5, 2020, www.newscientist.com/article/mg16722504-300-sick-to-death.

17 Debora MacKenzie, "Plague on a national icon," *New Scientist*, October 26, 2002, www.newscientist.com/article/mg17623661-100-plague-on-a-national-icon.

18 Lee Berger et al., "Chytridiomycosis causes amphibian mortality associated with population declines in the rain forests of Australia and Central America," *Proceedings of the National Academy of Sciences* 95, no. 15 (July 1998): 9031–9036, doi.org/10.1073/pnas.95.15.9031.

19 Kate E. Jones et al., "Global trends in emerging infectious diseases," *Nature* 451, no. 7181 (2008): 990–993, doi.org/10.1038/nature06536.

20 Almudena Marí Saéz et al., "Investigating the zoonotic origin of the West African Ebola epidemic," *EMBO Molecular Medicine* 7, no. 1 (January 2015), doi.org/10.15252/emmm.201404792.

21 Paul Nuki and Alanna Shaik, "Scientists put on alert for deadly new pathogen—'Disease X,'" *Telegraph*, March 10, 2018, www.telegraph.co.uk/global-health/science-and-disease/world-health-organization-issues-alert-disease-x.

22 "Factsheet about Crimean-Congo haemorrhagic fever," European Centre for Disease Prevention and Control (EU), www.ecdc.europa.eu/en/crimean-congo-haemorrhagic-fever/facts/factsheet.

23 Ana Negredo et al., "Survey of Crimean-Congo hemorrhagic fever enzootic focus, Spain, 2011–2015," *Emerging Infectious Diseases* 25, no. 6 (June 2019): 1177–1184, doi.org/10.3201/eid2506.180877.

24 Debora MacKenzie, "New killer virus makes an appearance," *New Scientist*, October 15, 2008, www.newscientist.com/article/mg20026783-200 -new-killer-virus-makes-an-appearance.

25 Shirlene Telmos Silva de Lima et al., "Fatal Outcome of Chikungunya Virus Infection in Brazil," *Clinical Infectious Diseases* ciaa1038 (August 2020), doi.org/10.1093/cid/ciaa1038.

26 Nuno Rodrigues Faria et al., "Zika virus in the Americas: Early epidemiological and genetic findings," *Science* 352, no. 6283 (April 2016): 345–349, doi.org/10.1126/science.aaf5036.

27 Ling Yuan et al., "A single mutation in the prM protein of Zika virus contributes to fetal microcephaly," *Science* 358, no. 6365 (November 2017): 933–936, doi.org/10.1126/science.aam7120.

28 Lai-Meng Looi, "Lessons from the Nipah virus outbreak in Malaysia," *Malaysian Journal of Pathology* 29, no. 2 (2007): 63–67, www.mjpath .org.my/2007.2/02Nipah_Virus_lessons.pdf.

29 Pragya D. Yadav et al., "Nipah virus sequences from humans and bats during Nipah outbreak, Kerala, India, 2018," *Emerging Infectious Diseases* 25, no. 5 (November 2020): 1003–1006, doi.org/10.3201 /eid2505.181076.

30 Chunyan Wang et al., "A human monoclonal antibody blocking SARS-CoV-2 Infection," *Nature Communications* 11, no. 2251 (May 12, 2020), doi.org/10.1101/2020.03.11.987958.

31 Olivier Pernet et al., "Evidence for henipavirus spillover into human populations in Africa," *Nature Communications* 5, no. 1 (November 2014), doi.org/10.1038/ncomms6342.

32 Debora MacKenzie, "World must get ready now for the next big health threat," *New Scientist*, December 15, 2015, www.newscientist.com /article/mg22830522-900-world-must-get-ready-now-for-the-next-big -health-threat.

33 WHO, "Ebola virus disease—Democratic Republic of the Congo," *Emergencies preparedness, response*, November 18, 2020, www.who.int /csr/don/18-november-2020-ebola-drc/en.

34 Debora MacKenzie, "Ebola rapidly evolves to be more transmissible and deadlier," *New Scientist*, November 3, 2016, www.newscientist

.com/article/2111311-ebola-rapidly-evolves-to-be-more-transmissible
-and-deadlier.

## CHAPTER 3: SARS, MERS — YOU CAN'T SAY WE WEREN'T WARNED

1 Nanshan Zhong and Guangqiao Zeng, "What we have learnt from SARS epidemics in China," *BMJ* 333, no. 7564 (August 2006): 389–391, doi.org/10.1136/bmj.333.7564.389.

2 ProMED-mail, "Pneumonia-China (Guangdong): RFI," *ProMED-mail Archive 20030210.0357*, February 10, 2003. Available at: www.promed mail.org.

3 ProMED-mail, "Pneumonia—China (Guangdong) (03)," *ProMED-mail Archive 20030214.039*, February 14, 2003. Available at: www.prome dmail.org.

4 ProMED-mail, "Pneumonia—China (Guangdong) (04)," *ProMED-mail Archive 20030219.0427*, February 19, 2003. Available at: www.prome dmail.org.

5 ProMED-mail, "Pneumonia—China (Guangdong) (06)," *ProMED-mail Archive 20030220.0447*, February 20, 2003. Available at: www.prome dmail.org.

6 Meredith Wadman, Jennifer Couzin-Frankel, Jocelyn Kaiser, and Catherine Matacic, "How does coronavirus kill? Clinicians trace a ferocious rampage through the body, from brain to toes," *Science*, April 17, 2020, www.sciencemag.org/news/2020/04/how-does-coronavirus-kill-clini cians-trace-ferocious-rampage-through-body-brain-toes.

7 Christian Kreuder-Sonnen, "China vs the WHO: A Behavioural Norm Conflict in the SARS Crisis," *International Affairs* 95, no. 3 (January 2019): 535–552, doi.org/10.1093/ia/iiz022.

8 Tim Brookes with Omar A. Khan, *Behind the Mask: How the World Survived SARS, the First Epidemic of the 21st Century* (Washington, DC: American Public Health Association, 2005), 195.

9 Yanzhong Huang, "The SARS epidemic and its aftermath in China: A political perspective," in *Learning from SARS: Preparing for the Next Disease Outbreak: Workshop Summary*, ed. Stacey Knobler et al. (Washington, DC: National Academies Press, 2004), www.ncbi.nlm.nih.gov/books/NBK 92479.

# Notes

10 Yanzhong Huang, "The SARS epidemic and its aftermath in China: A political perspective."

11 Gordon Gauchat, "Politicization of science in the public sphere: A study of public trust in the United States, 1974 to 2010," *American Sociological Review* 77, no. 2 (2012): 167–187, doi.org/10.1177/0003122412438225.

12 World Health Organization, *The World Health Report 2003: Shaping the Future* (Geneva, Switzerland: WHO, 2003), www.who.int/whr/2003/en.

13 Mark Henderson, "End of SARS as a deadly threat," *Times of London*, February 21, 2009, www.thetimes.co.uk/article/end-of-sars-as-a-deadly -threat-nz3ll7tqzsz.

14 L. F. Wang and B. T. Eaton, "Bats, civets and the emergence of SARS," *Current Topics in Microbiology and Immunology Wildlife and Emerging Zoonotic Diseases: The Biology, Circumstances and Consequences of Cross-Species Transmission* 315 (2007): 325–344, doi.org/10.1007/978-3-540-70962-6_13.

15 Zhang Feng, "Does SARS virus still exist in the wild?," *China Daily*, February 23, 2005, www.chinadaily.com.cn/english/doc/2005-02/23/con tent_418481.htm.

16 Nanshan Zhong and Guangqiao Zeng, "What we have learnt from SARS epidemics in China."

17 ProMED-mail, "Novel coronavirus—Saudi Arabia: human isolate," *ProMED-mail Archive 20120920.1302733*, September 20, 2012, www. promedmail.org.

18 Debora MacKenzie, "Threatwatch: Find the germs, don't sack the messenger," *New Scientist*, October 24, 2012, www.newscientist.com /article/dn22417-threatwatch-find-the-germs-dont-sack-the-messenger.

19 Kate Kelland, "Special Report—Saudi Arabia takes heat for spread of MERS virus," Reuters, May 22, 2014, uk.reuters.com/article /uk-saudi-mers-special-report/special-report-saudi-arabia-takes-heat -for-spread-of-mers-virus-idUKKBN0E207Z20140522.

20 Christl A. Donnelly et al., "Worldwide reduction in MERS cases and deaths since 2016," *Emerging Infectious Diseases* 25, no. 9 (September 2019): 1758–1760, doi.org/10.3201/eid2509.190143.

21 Debora MacKenzie, "Secrets and Lies in Europe," *New Scientist*, May 3, 1997, www.newscientist.com/article/mg15420802-300-secrets-and-lies -in-europe.

# Notes

## CHAPTER 4: DON'T BLAME THE BATS

1 World Health Organization, "Global hepatitis report, 2017," 2017, apps.who.int/iris/handle/10665/255016.

2 Marc T. Valitutto et al., "Detection of novel coronaviruses in bats in Myanmar," *PLoS One* 15, no. 4 (April 2020): e0230802, doi.org/10.1371/journal.pone.0230802.

3 Smriti Mallapaty, "Coronaviruses closely related to the pandemic virus discovered in Japan and Cambodia," *Nature* 588 (November 2020): 15–16, doi.org/10.1038/d41586-020-03217-0.

4 Simon J. Anthony et al., "Global patterns in coronavirus diversity," *Virus Evolution* 3, no. 1 (January 2017), doi.org/10.1093/ve/vex012.

5 Anthony King, "Super bats: What doesn't kill them, could make us stronger," *New Scientist*, February 10, 2016, www.newscientist.com/article/2076598-super-bats-what-doesnt-kill-them-could-make-us-stronger.

6 Kevin J. Olival et al., "Host and Viral Traits Predict Zoonotic Spillover from Mammals," *Nature* 546, no. 7660 (June 2017): 646–650, doi.org/10.1038/nature22975.

7 Wendong Li et al., "Bats are natural reservoirs of SARS-like coronaviruses," *Science* 310, no. 5748 (October 2005): 676–679, doi.org/10.1126/science.1118391.

8 Xing-Yi Ge et al., "Isolation and characterization of a bat SARS-like coronavirus that uses the ACE2 receptor," *Nature* 503, no. 7477 (October 2013): 535–538, doi.org/10.1038/nature12711.

9 Ben Hu et al., "Discovery of a rich gene pool of bat SARS-related coronaviruses provides new insights into the origin of SARS coronavirus," *PLoS Pathogens* 13, no. 11 (November 2017), doi.org/10.1371/journal.ppat.1006698.

10 Menachery et al., "A SARS-like cluster of circulating bat coronaviruses shows potential for human emergence."

11 Vineet D. Menachery et al., "SARS-like WIV1-CoV poised for human emergence," *Proceedings of the National Academy of Sciences* 113, no. 11 (March 2016): 3048–3053, doi.org/10.1073/pnas.1517719113.

12 Debora MacKenzie, "Plague! How to prepare for the next pandemic," *New Scientist*, February 22, 2017, www.newscientist.com/article

/mg23331140-400-plague-how-to-prepare-for-the-next-pandemic/#ixz
z6KMAMFWDf.

13 Ning Wang et al., "Serological evidence of bat SARS-related coronavirus infection in humans, China." *Virologica Sinica* 33, no. 1 (February
2018): 104–107, doi.org/10.1007/s12250-018-0012-7.

14 Yi Fan et al., "Bat coronaviruses in China," *Viruses* 11, no. 3 (March
2019): 210, doi.org/10.3390/v11030210.

15 Peng Zhou et al., "A pneumonia outbreak associated with a new
coronavirus of probable bat origin," *Nature* 579, no. 7798 (February 2020):
270–273, doi.org/10.1038/s41586-020-2012-7.

16 Public Health England, "Briefing Note—Serial number 2018/054,"
October 11, 2018, www.england.nhs.uk/south/wp-content/uploads/sites/6
/2018/11/briefing-note-bat-rabies-20181011.pdf.

17 Exotic Disease Control Team (UK government), "Rabies control
strategy for Great Britain," June 2018, rev. August 2019, assets.publishing
.service.gov.uk/government/uploads/system/uploads/attachment
_data/file/831362/rabies-control-strategy-aug2019a.pdf.

18 Tommy Tsan-Yuk Lam et al., "Identifying SARS-CoV-2 related coronaviruses in Malayan pangolins," *Nature*, March 26, 2020, doi.org/10.1038
/s41586-020-2169-0.

19 Maciej F. Boni et al., "Evolutionary origins of the SARS-CoV-2 sar-
becovirus lineage responsible for the COVID-19 pandemic," *Nature Microbiology* 5 (July 2020): 1408–1417, doi.org/10.1101/2020.03.30.015008.

20 Yiran Wu and Suwen Zhao, "Furin cleavage sites naturally occur
in coronaviruses," *Stem Cell Research* 50 (January 2021): 102115, doi
.org/10.1016/j.scr.2020.102115.

21 Hong Zhou et al., "Identification of novel bat coronaviruses sheds
light on the evolutionary origins of SARS-CoV-2 and related viruses,"
BioRxiv, March 8. 2021, https://www.biorxiv.org/content/10.1101/2021.0
3.08.434390v1.

22 Tang Ailin, Chen Yifan, and Matthew Walsh, "China's epidemic-
inspired wildlife ban has had big economic costs," *Nikkei Asia*, August 13, 2020,
asia.nikkei.com/Spotlight/Caixin/China-s-epidemic-inspired-wildlife
-ban-has-had-big-economic-costs.

# Notes

23 Yew Lun Tian and David Stanway, "China's fur farms see opportunity as countries cull mink over coronavirus fears," Reuters, December 3, 2020, www.reuters.com/article/us-health-coronavirus-china-mink-idUSKBN 28D0PV.

24 Tammy Mildenstein, Iroro Tanshi, and Paul A. Racey, "Exploitation of bats for bushmeat and medicine," in *Bats in the Anthropocene: Conservation of Bats in a Changing World*, ed. Christian C. Voigt and Tigga Kingston (Cham, Switzerland: Springer Open, 2016), doi.org/10.1007/978-3-319-25220-9_12.

25 "Ye Ming Sha, bat feces, bat dung, bat guano," Best Plant, www.best plant.shop/products/ye-ming-sha-bat-feces-bat-dung-bat-guano.

26 Chun-Han Zhu, *Clinical Handbook of Chinese Prepared Medicines* (Brookline, MA: Paradigm, 1989), 179.

27 Peter Borten, "Chinese herbs," chineseherbinfo.com/ye-ming-sha -bat-feces.

28 John Xie, "Did coronavirus come from the bat guano trade?," *Voice of America*, May 11, 2020, www.voanews.com/covid-19-pandemic /did-coronavirus-come-bat-guano-trade.

29 "Why China's traditional medicine boom is dangerous," *The Economist*, September 2, 2017, www.economist.com/china/2017/09/01/why -chinas-traditional-medicine-boom-is-dangerous.

30 National Health Commission of the People's Republic of China, "Number of certified TCM professionals in China increases 7.2% in 2019," June 7, 2020, en.nhc.gov.cn/2020-06/07/c_80664.htm#:~:text=Number%20 of%20certified%20TCM%20professionals%20in%20China%20 increases%207.2%25%20in%202019,-Updated%3A%202020%2D06&text =BEIJING%20%2D%20The%20number%20of%20certified,by%20 China's%20National%20Health%20Commission.

31 Francesca Colavita et al., "SARS-CoV-2 isolation from ocular secretions of a patient with COVID-19 in Italy with prolonged viral RNA detection," *Annals of Internal Medicine* [Epub ahead of print April 17, 2020], doi.org/10.7326/M20-1176.

32 Kenrie P. Y. Hui, "Tropism, replication competence, and innate immune responses of the coronavirus SARS-CoV-2 in human respiratory tract and conjunctiva: An analysis in ex-vivo and in-vitro cultures," *The Lancet Respiratory Medicine*, May 7, 2020, doi.org/10.1016/S2213-2600(20)30193-4.

# Notes

33 T. M. Wassenaar and Y. Zou, "2019_nCoV/SARS-CoV-2: Rapid Classification of Betacoronaviruses and Identification of Traditional Chinese Medicine as Potential Origin of Zoonotic Coronaviruses," *Letters in Applied Microbiology* 70, no. 5 (February 2020): 342–348, doi.org/10.1111/lam.13285.

34 Supaporn Wacharapluesadee et al., "Group C Betacoronavirus in Bat Guano Fertilizer, Thailand," *Emerging Infectious Diseases* 19, no. 8 (August 2013): 1349–1352, doi.org/10.3201/eid1908.130119.

35 "China to close all live poultry markets gradually," *Xinhua*, July 3, 2020, www.xinhuanet.com/english/2020-07/03/c_139186436.htm.

36 John Garrick and Yan Bennett, "How China is controlling the COVID origins narrative—silencing critics and locking up dissenters," *The Conversation*, January 13, 2021, theconversation.com/how-china-is-con trolling-the-covid-origins-narrative-silencing-critics-and-locking-up-dis senters-152751.

37 Forestry Administration of Guangdong Province, "穿山甲、蝙蝠粪便将不再入药！2020版药典已将其删除！" June 11, 2020.

38 Jani Actman, "Traditional Chinese medicine and wildlife," *National Geographic*, February 7, 2019, www.nationalgeographic.com/animals /reference/traditional-chinese-medicine.

39 Duke-NUS Graduate Medical School, "Researchers find genetic link between bats' ability to fly and viral immunity," Duke Global Health Institute, December 20, 2012, globalhealth.duke.edu/news/researchers -find-genetic-link-between-bats-ability-fly-and-viral-immunity.

40 Jiazheng Xie et al., "Dampened STING-dependent interferon activation in bats," *Cell Host & Microbe* 23, no. 3 (March 2018), doi.org /10.1016/j.chom.2018.01.006.

41 Cara E. Brook et al., "Accelerated viral dynamics in bat cell lines, with implications for zoonotic emergence," *eLife* (February 2020), doi .org/10.7554/eLife.48401.

42 Thelma Gómez Durán, "En defensa de los murciélagos: Resistentes a los virus, pero no a los humanos," *Mongabay Latam*, March 31, 2020, es.mong abay.com/2020/03/coronavirus-murcielagos-humanos-virus-covid-19.

43 Yash Goyal, "More than 150 bats killed in Rajasthan owing to fear of COVID-19 spread," *Tribune India*, May 7, 2020, www.tribuneindia .com/news/nation/more-than-150-bats-killed-in-rajasthan-owing-to-fe

ar-of-covid-19-spread-81668?fbclid=IwAR0WcG8b_EIRVDOJCYTi_jm
VNiFrCduH_JRzNVUu_2_EBmLl51LTJxQ9IbY.

44 Rhiannon Tuffield and Sarah Maunder, "Vulnerable grey-headed flying foxes shot and bashed in regional Victoria," *ABC Shepparton*, March 9, 2020, www.abc.net.au/news/2020-03-09/bat-deaths-grey-headed-flying-fox/12039936.

45 Newsflare, "Hundreds of bats burned in Indonesia in bid to prevent coronavirus spread," *Yahoo! News*, March 16, 2020, news.yahoo.com/hundreds-bats-burned-indonesia-bid-150000233.html.

46 R. Rocha et al., "Bat Conservation and Zoonotic Disease Risk: a Research Agenda to Prevent Misguided Persecution in the Aftermath of COVID-19," *Animal Conservation* (2020):doi.org/10.1111/acv.12636.

47 J. G. Boyles et al., "Economic Importance of Bats in Agriculture," *Science* 332, no. 6025 (April 2011): 41–42, doi.org/10.1126/science.1201366.

48 Alexandra Zimmerman, Ewan Macdonald, and Tigga Kingston, "Why Mauritius is culling an endangered fruit bat that exists nowhere else," *The Conversation*, November 26, 2020, theconversation.com/why-mauritius-is-culling-an-endangered-fruit-bat-that-exists-nowhere-else-150567.

49 Boyles et al., "Economic Importance of Bats in Agriculture."

50 "Bat Conservation International bats and disease position statement," Bats & Human Health, Bat Conservation International, www.batcon.org/resources/for-specific-issues/bats-human-health.

51 Charles H. Calisher et al., "Bats: Important reservoir hosts of emerging viruses," *Clinical Microbiology Reviews* 19, no. 3 (July 2006): 531–545, doi.org/10.1128/cmr.00017-06.

52 Raina K. Plowright et al., "Ecological dynamics of emerging bat virus spillover," *Proceedings of the Royal Society B: Biological Sciences* 282, no. 1798 (January 7, 2015): 2014–2124, doi.org/10.1098/rspb.2014.2124.

53 Raina K. Plowright et al., "Reproduction and nutritional stress are risk factors for Hendra virus infection in little red flying foxes (*Pteropus scapulatus*)," *Proceedings of the Royal Society B: Biological Sciences* 275, no. 1636 (January 2008): 861–869, doi.org/10.1098/rspb.2007.1260.

54 Daniel G. Streicker et al., "Ecological and anthropogenic drivers of rabies exposure in vampire bats: Implications for transmission and control," *Proceedings of the Royal Society B: Biological Sciences* 279, no. 1742 (August 2012): 3384–3392, doi.org/10.1098/rspb.2012.0538.

# Notes

55 R. Rocha et al., "Bat Conservation and Zoonotic Disease Risk."

56 David M. Cutler and Lawrence H. Summers, "The COVID-19 pandemic and the $16 trillion virus," *JAMA* 324, no. 15 (October 2020): 1495, doi.org/10.1001/jama.2020.19759.

57 Andrew P. Dobson et al., "Ecology and economics for pandemic prevention," *Science* 369, no. 6052 (July 2020): 379–381, doi.org/10.1126/science.abc3189.

58 Intergovernmental Science-Policy Platform on Biodiversity and Ecosystem Services, "IPBES #PandemicsReport: Escaping the 'Era of Pandemics,'" 2020, www.ipbes.net/pandemics-report-complete-digital-version.

59 Mark Terry, "NIH awards EcoHealth Alliance $7.5 million grant despite political furor," *BioSpace*, August 28, 2020, www.biospace.com/article/1nih-awards-ecohealth-alliance-7-5-million-grant-despite-political-furor.

60 Andrew Silver and David Cyranoski, "China is tightening its grip on coronavirus research," *Nature* 580, no. 7804 (April 2020): 439–440, doi.org/10.1038/d41586-020-01108-y.

## CHAPTER 5: WASN'T THE PANDEMIC SUPPOSED TO BE FLU?

1 Ron A. M. Fouchier et al., "Koch's postulates fulfilled for SARS virus," *Nature* 423 (May 2003): 240, doi.org/10.1038/423240a.

2 Colin A. Russell et al., "The global circulation of seasonal influenza A (H3N2) viruses," *Science* 320, no. 5874 (April 2008), doi.org/10.1126/science.1154137.

3 Debora MacKenzie, "Jab in the dark: Why we don't have a universal flu vaccine," *New Scientist*, January 2, 2018, www.newscientist.com/article/2156915-jab-in-the-dark-why-we-dont-have-a-universal-flu-vaccine.

4 R. J. Webby et al., "Multiple lineages of antigenically and genetically diverse influenza A virus co-circulate in the United States swine population," *Virus Research* 103, no. 1–2 (July 2004): 67–73, doi.org/10.1016/j.virusres.2004.02.015.

5 Shanta Zimmer and Donald Burke, "Historical Perspective—Emergence of Influenza A (H1N1) Viruses," New England Journal of Medicine 2009; 361:279-285, https://www.nejm.org/doi/10.1056/NEJMra0904322?url_ver=Z39.88-2003&rfr_id=ori%3Arid%3Acrossref.org&rfr_dat=cr_pub++0www.ncbi.nlm.nih.gov.

## Notes

6 Michelle Rozo and Gigi Kwik Gronvall, "The reemergent 1977 H1N1 strain and the gain-of-function debate," *MBio* 6, no. 4 (August 2015): e01013–e01015, doi.org/10.1128/mbio.01013-15.

7 Laura MacInnis and Stephanie Nebehay, "WHO warns flu pandemic imminent," Reuters, April 28, 2009, www.reu ters.com/article/us-flu/who -warns-flu-pandemic-imminent-idUSTRE 53N22820090429.

8 "FAO acts over H1N1 human crisis," Food and Agriculture Organization of the United Nations, April 27, 2009, www.fao.org/news/story/en /item/13002/icode.

9 "WHO pandemic declaration," Centers for Disease Control and Prevention, www.cdc.gov/h1n1flu/who.

10 "Genes could be key to new Covid-19 treatments," University of Edinburgh, 2020, www.ed.ac.uk/news/2020/genes-could-be-key-to-new-covid -19-treatments.

11 Richard Knox, "Flu pandemic much milder than expected," *NPR Morning Edition*, December 8, 2009, www.npr.org/templates/story/story .php?storyId=121184706.

12 "COVID-19 pandemic just started, hard to see end: Chinese epidemiologist," *Global Times*, March 24, 2020, www.globaltimes.cn/con tent/1183619.shtml.

13 Public Health England, "Pandemic Influenza Response Plan 2014," August 2014, assets.publishing.service.gov.uk/government/uploads/system /uploads/attachment_data/file/344695/PI_Response_Plan_13_Aug.pdf[0].

14 Angela N. Cauthen et al., "Continued circulation in China of highly pathogenic avian influenza viruses encoding the hemagglutinin gene associated with the 1997 H5N1 outbreak in poultry and humans," *Journal of Virology* 74, no. 14 (July 2000): 6592–6599, doi.org/10.1128 /jvi.74.14.6592-6599.2000.

15 Y. Guan et al., "Emergence of multiple genotypes of H5N1 avian influenza viruses in Hong Kong SAR," *Proceedings of the National Academy of Sciences* 99, no. 13 (June 2002): 8950–8955, doi.org/10.1073/pnas .132268999.

16 Debora MacKenzie, "Bird flu outbreak started a year ago," *New Scientist*, January 28, 2004, www.newscientist.com/article/dn4614-bird-flu -outbreak-started-a-year-ago.

## Notes

17 Reuters, "China denies bird flu cover-up," *CNN International*, January 29, 2004, edition.cnn.com/2004/WORLD/asiapcf/01/28/bird.flu.china.reut.

18 Oliver August, "China covers up again on outbreak," *The Times*, February 2, 2004, www.thetimes.co.uk/article/china-covers-up-again-on-out break-hntz3rp3rgj.

19 H. Chen et al., "Establishment of multiple sublineages of H5N1 influenza virus in Asia: Implications for pandemic control," *Proceedings of the National Academy of Sciences* 103, no. 8 (February 2006): 2845–2850, doi.org/10.1073/pnas.0511120103.

20 H. Chen et al., "H5N1 virus outbreak in migratory waterfowl," *Nature* 436, no. 7048 (July 2005): 191–192, doi.org/10.1038/nature03974.

21 Debora MacKenzie, "China denies bird flu research findings," *New Scientist*, July 13, 2005, www.newscientist.com/article/mg18725083 -500-china-denies-bird-flu-research-findings.

22 European Centre for Disease Prevention and Control (EU), "Avian influenza overview December 2020—February 2021," February 26, 2021, www.ecdc.europa.eu/en/publications-data/avian-influenza-overview -december-2020-february-2021.

23 WHO, "Human infection with avian influenza A (H5N8)—the Russian Federation," Disease Outbreak News, February 28, 2021, www.who .int/csr/don/26-feb-2021-influenza-a-russian-federation/en.

24 Jacqueline King et al., "Novel HPAIV H5N8 Reassortant (Clade 2.3.4.4b) Detected in Germany," *Viruses* 12, no. 3 (April 2020): 281, doi .org/10.3390/v12030281.

25 Anni McLeod et al., "Economic and social impacts of avian influenza," FAO Emergency Centre for Transboundary Animal Diseases Operations (ECTAD), November 2005, www.fao.org/avianflu/documents/Economic -and-social-impacts-of-avian-influenza-Geneva.pdf.

26 Public Health England, "Risk assessment of avian influenza A(H7N9)—eighth update," January 8, 2020, www.gov.uk/government/publi cations/avian-influenza-a-h7n9-public-health-england-risk-assessment /risk-assessment-of-avian-influenza-ah7n9-sixth-update.

27 S. Herfst et al., "Airborne transmission of influenza A/H5N1 virus between ferrets," *Science* 336, no. 6088 (June 21, 2012): 1534–1541, doi .org/10.1126/science.1213362.

28 Masaki Imai et al., "A highly pathogenic avian H7N9 influenza virus isolated from a human is lethal in some ferrets infected via respiratory droplets," *Cell Host & Microbe* 22, no. 5 (November 2017), doi .org/10.1016/j.chom.2017.09.008.

29 Anthony S. Fauci, "Research on highly pathogenic H5N1 influenza virus: The way forward," *MBio* 3, no. 5 (October 2012), doi.org/10.1128 /mbio.00359-12.

30 National Institutes of Health, "Notice announcing the removal of the funding pause for gain-of-function research projects," December 19, 2017, grants.nih.gov/grants/guide/notice-files/NOT-OD-17-071.html.

31 Peter Daszak (EcoHealth Alliance), "Understanding the risk of bat coronavirus emergence," Project Number: 2R01AI110964-06, NIH Research Portfolio Online Reporting Tools (RePORT), projectreporter.nih .gov/project_info_description.cfm?aid=9819304&icde=49645421.

32 EcoHealth Alliance, "Regarding NIH termination of coronavirus research funding," April 2020, www.ecohealthalliance.org/2020/04 /regarding-nih-termination-of-coronavirus-research-funding.

33 Honglei Sun et al., "Prevalent Eurasian avian-like H1N1 swine influenza virus with 2009 pandemic viral genes facilitating human infection," *PNAS* 117, no. 29 (June 2020): 17204–17210, doi.org/10.1073 /pnas.1921186117.

34 Dinah Henritzi et al., "Surveillance of European domestic pig populations identifies an emerging reservoir of potentially zoonotic swine influenza A viruses," *Cell Host & Microbe* 28, no. 4 (July 2020): 614–627.E6, doi.org/10.1016/j.chom.2020.07.006.

## CHAPTER 6: SO WHAT DO WE DO ABOUT DISEASE?

1 Bill Gates, "Innovation for pandemics," *The New England Journal of Medicine* 378 (May 2018): 2057–2060, doi.org/0.1056/NEJMp1806283. Remarks originally delivered as the Shattuck Lecture for the Massachusetts Medical Society on April 27, 2018.

2 Christopher Kirchhoff, "Memorandum for Ambassador Susan E. Rice, Subject: NSC Lessons Learned Study on Ebola," National Security Council, White House, July 11, 2016, assets.documentcloud.org/docu ments/6817684/NSC-Ebola-Lessons-Learend-Report-FINAL-8-28-16.pdf.

# Notes

3 Christopher Kirchhoff, "Ebola should have immunized the United States to the coronavirus," *Foreign Affairs*, March 28, 2020, www.foreig naffairs.com/articles/2020-03-28/ebola-should-have-immunized-united -states-coronavirus.

4 Tedros Ghebreyesus, "WHO director-general's opening remarks at the media briefing on COVID-19," March 11, 2020, www.who.int/dg /speeches/detail/who-director-general-s-opening-remarks-at-the-me dia-briefing-on-covid-1911-march-2020.

5 The Independent Panel, "Second Report."

6 Yasmeen Abutaleb, Josh Dawsey, Ellen Nakashima, and Greg Miller, "The U.S. was beset by denial and dysfunction as the coronavi-rus raged," *Washington Post*, April 4, 2020, www.washingtonpost.com /national-security/2020/04/04/coronavirus-government-dysfunction.

7 Global Preparedness Monitoring Board, "A world at risk: Annual report on global preparedness for health emergencies," September 2019, apps.who.int/gpmb/assets/annual_report/GPMB_Annual_Report _English.pdf.

8 United Nations, High-Level Panel on the Global Response to Health, "Protecting humanity from future health crises: Report of the High-Level Panel on the Global Response to Health Crises," February 2016, www .un.org/ga/search/view_doc.asp?symbol=A/70/723.

9 "UK forms global infection response team," *BBC News*, November 1, 2016, www.bbc.com/news/health-37827388.

10 Global Preparedness Monitoring Board, "A world at risk: Annual report on global preparedness for health emergencies."

11 G20, "G20 leaders' statement, extraordinary G20 leaders' summit statement on COVID-19," March 26, 2020, g20.org/en/media/Documents /G20_Extraordinary%20G20%20Leaders%E2%80%99%20Summit_State ment_EN%20(3).pdf.

12 "Full text of the G20 leaders final communique at the end of the G20 Riyadh Summit," *Al Arabiya News*, November 23, 2020, english.alara biya.net/News/gulf/2020/11/22/Full-text-of-the-G20-leaders-final-com munique-at-the-end-of-the-G20-Riyadh-Summit.

13 The Johns Hopkins Center for Health Security, "The character-istics of pandemic pathogens," 2018, www.centerforhealthsecurity.org

/our-work/pubs_archive/pubs-pdfs/2018/180510-pandemic-pathogens
-report.pdf.

14 Pekar et al., "Timing the SARS-CoV-2 Index Case in Hubei Province."

15 Katherine E. Graham et al., "SARS-CoV-2 RNA in Wastewater Settled Solids Is Associated with COVID-19 Cases in a Large Urban Sewer-
shed," *Environmental Science & Technology* 55 (2021): 488–498, doi.org
/10.1021/acs.est.0c06191.

16 Debora MacKenzie, "Germ detectors: Unmasking our microbial
foes," *New Scientist*, August 17, 2011, www.newscientist.com/article/mg
21128262-400-germ-detectors-unmasking-our-microbial-foes.

17 National Institutes of Health, "NIH to support radical approaches to
nationwide COVID-19 testing and surveillance," December 21, 2020, www
.nih.gov/news-events/news-releases/nih-support-radical-approaches
-nationwide-covid-19-testing-surveillance.

18 Cormac Sheridan, "COVID-19 Spurs Wave of Innovative Diagnostics," *Nature Biotechnology* 38, no. 7 (July 2020): 769–772, doi.org/10.1038
/s41587-020-0597-x.

19 Edward C. Holmes, Andrew Rambaut, and Kristian G. Andersen,
"Pandemics: Spend on surveillance, not prediction," *Nature* 558, no. 7709
(June 7, 2018): 180–182, doi.org/10.1038/d41586-018-05373-w.

20 "Our Approach," Global Virome Project, www.globalviromeproject
.org/our-approach.

21 World Health Organization, "International Health Regulations, 2nd
edition," 2005, www.who.int/ihr/9789241596664/en.

22 Sarah Boseley, "World Health Organisation 'intentionally delayed
declaring Ebola emergency,'" *Guardian*, March 20, 2015, www.theguard
ian.com/world/2015/mar/20/ebola-emergency-guinea-epidemic-who.

23 Global Health Security Index, "2019 GHS Index," 2019, www.ghsin
dex.org/wp-content/uploads/2019/10/2019-Global-Health-Security-In
dex.pdf.

24 Najmul Haider et al., "The Global Health Security index and Joint
External Evaluation score for health preparedness are not correlated
with countries' COVID-19 detection response time and mortality outcome," *Epidemiology & Infection* 148 (2020): e210, doi.org/10.1017
/S0950268820002046.

## Notes

25 Ngaire Woods, "The Brutal Governance Lessons of 2020," *Project Syndicate*, December 29, 2020, www.project-syndicate.org/commentary /governance-lessons-from-covid19-explaining-us-uk-failure-by-ngaire -woods-2020-12.

26 Paul Nuki, "The six crucial pandemic lessons the government ignored," *The Telegraph*, October 25, 2020, www.telegraph.co.uk/global-health /science-and-disease/six-crucial-pandemic-lessons-government-ignored.

27 David Pegg, "Official report that said UK was not prepared for pandemic is published," *The Guardian*, October 22, 2020, www.theguardian .com/world/2020/oct/22/official-report-exercise-cygnus-uk-was-not-pre pared-for-pandemic-is-published.

28 David E. Sanger, Eric Lipton, Eileen Sullivan, and Michael Crowley, "Before virus outbreak, a cascade of warnings went unheeded," *New York Times*, March 22, 2020, www.nytimes.com/2020/03/19/us/politics /trump-coronavirus-outbreak.html.

29 Lawrence O. Gostin and Eric A. Friedman, "Ebola: A crisis in global health leadership," *The Lancet* 384, no. 9951 (October 2014): 1323–1325, doi.org/10.1016/s0140-6736(14)61791-8.

30 Global Preparedness Monitoring Board, "A world in disorder," September 14, 2020, apps.who.int/gpmb/assets/annual_report/2020/GPMB _2020_AR_EN_WEB.pdf.

31 The Independent Panel, "Second Report."

32 The White House, "National Security Memorandum on United States Global Leadership to Strengthen the International COVID-19 Response and to Advance Global Health Security and Biological Preparedness," January 21, 2021, www.whitehouse.gov/briefing-room /statements-releases/2021/01/21/national-security-directive-uni ted-states-global-leadership-to-strengthen-the-international-covid-19 -response-and-to-advance-global-health-security-and-biological -preparedness.

33 Caitlin Rivers and Dylan George, "How to forecast outbreaks and pandemics," *Foreign Affairs*, June 29, 2020, www.foreignaffairs.com/articles /united-states/2020-06-29/how-forecast-outbreaks-and-pandemics.

34 Scott Gottlieb et al., "National coronavirus response: A road map to reopening," American Enterprise Institute, March 29, 2020, www.aei.org

/research-products/report/national-coronavirus-response-a-road-map
-to-reopening.

35 The Independent Panel, "Second Report."

36 Martin Furmanski, "Laboratory escapes and 'self-fulfilling proph-
ecy' epidemics," Center for Arms Control and Non-Proliferation, February
17, 2014, https://armscontrolcenter.org/wp-content/uploads/2016/02/
Escaped-Viruses-final-2-17-14-copy.pdf.

37 Debora MacKenzie, "US develops lethal new viruses," *New Scientist*,
October 29, 2003, www.newscientist.com/article/dn4318-us-develops-le
thal-new-viruses.

38 Kristian G. Andersen et al., "The proximal origin of SARS-CoV-2,"
*Nature Medicine* 26, no. 4 (March 17, 2020): 450–452, doi.org/10.1038
/s41591-020-0820-9.

39 Edward C. Holmes et al., "Spike protein sequences of Cambo-
dian, Thai and Japanese bat sarbecoviruses provide insights into the
natural evolution of the Receptor Binding Domain and S1/S2 cleavage site," *Virological*, February 21, 2021, virological.org/t/spike-protein
-sequences-of-cambodian-thai-and-japanese-bat-sarbecoviruses-pro
vide-insights-into-the-natural-evolution-of-the-receptor-binding
-domain-and-s1-s2-cleavage-site/622.

40 Benjamin Wallace-Wells, "The Sudden Rise of the Coronavirus
Lab-Leak Theory," *New Yorker*, May 27, 2021, https://www.newyorker.com
/news/annals-of-inquiry/the-sudden-rise-of-the-coronavirus-lab-leak-theory.

41 Jon Cohen, "Wuhan coronavirus hunter Shi Zhengli speaks out,"
*Science* 369, no. 6503 (July 2020): 487–488, doi.org/10.1126/science.369
.6503.487.

42 Michael Pompeo, "Ensuring a Transparent, Thorough Investigation of COVID-19's Origin," US Department of State, January 15, 2021,
2017-2021.state.gov/ensuring-a-transparent-thorough-investigation-of-covid
-19s-origin/index.html.

43 Botao Xiao et al., "The possible origins of 2019-ncov coronavirus,"
Researchgate preprint, February 2020, img-prod.tgcom24.mediaset.it/images
/2020/02/16/114720192-5eb8307f-017c-4075-a697-348628da0204.pdf.

44 www.xinhuanet.com/local/2017-05/03/c_1120909064_2.htm.

## Notes

45 Wen-Ping Guo et al., "Phylogeny and Origins of Hantaviruses Harbored by Bats, Insectivores, and Rodents," *PLoS Pathogens*, February 7, 2013, doi.org/10.1371/journal.ppat.1003159.

46 "WHO-Convened Global Study."

47 Jon Cohen, "Wuhan Coronavirus Hunter."

48 Charles Calisher et al., "Statement in support of the scientists, public health professionals, and medical professionals of China combatting COVID-19," *The Lancet* 395, no. 10226 (February 2020), doi.org/10.1016/s0140-6736(20)30418-9.

49 Albert D. M. E. Osterhaus et al., "Make science evolve into a One Health approach to improve health and security: A white paper," *One Health Outlook* 2, no. 6 (2020), doi.org/10.1186/s42522-019-0009-7.

50 Debora MacKenzie, "Is the common cold becoming a killer?," *New Scientist*, September 3, 2008, www.newscientist.com/article/mg19926721-900-is-the-common-cold-becoming-a-killer.

51 Kenneth A. Mclean et al., "The 2015 global production capacity of seasonal and pandemic influenza vaccine," *Vaccine* 34, no. 45 (October 2016): 5410–5413, doi.org/10.1016/j.vaccine.2016.08.019.

52 BiondVax Pharmaceuticals Ltd., "BiondVax Announces Topline Results from Phase 3 Clinical Trial of the M-001 Universal Influenza Vaccine Candidate," October 23, 2020, www.biondvax.com/2020/10/biondvax-announces-topline-results-from-phase-3-clinical-trial-of-the-m-001-universal-influenza-vaccine-candidate.

53 WHO, "UNICEF, WHO, IFRC and MSF announce the establishment of a global Ebola vaccine stockpile," January 12, 2021, www.who.int/news/item/12-01-2021-unicef-who-ifrc-and-msf-announce-the-establishment-of-a-global-ebola-vaccine-stockpile.

54 Press release, "MSF to wealthy countries: Don't block and ruin the potential of a landmark waiver on monopolies during the pandemic," Geneva, February 3, 2021, msfaccess.org/msf-wealthy-countries-dont-block-and-ruin-potential-landmark-waiver-monopolies-during-pandemic.

55 Scott Gottlieb et al., "National coronavirus response: A road map to reopening."

## Notes

56 Coalition for Epidemic Preparedness Innovations, "Landmark global collaboration launched to defeat COVID-19 pandemic," April 24, 2020, cepi.net/news_cepi/landmark-global-collaboration-launched-to-de feat-covid-19-pandemic.

57 Lisa Schnirring, "WHO director calls for easing of vaccine hurdles for developing nations," CIDRAP News, University of Minnesota, January 8, 2021, www.cidrap.umn.edu/news-perspective/2021/01/who-direc tor-calls-easing-vaccine-hurdles-developing-nations.

58 Mark McClellan et al., *Reducing Global COVID Vaccine Shortages: New Research and Recommendations for US Leadership*, April 15, 2001, https://healthpolicy.duke.edu/sites/default/files/2021-04/US%20Leader ship%20for%20Global%20Vaccines_1.pdf.

59 Marco Haffner et al., "The global economic cost of COVID-19 vaccine nationalism," RAND Corporation Research Briefs, 2020, doi.org /10.7249/RBA769-1.

60 International Chamber of Commerce, "Key findings and implications of the new study: THE ECONOMIC CASE FOR GLOBAL VACCINA-TION: An Epidemiological Model with International Production Networks (Cem Çakmaklı et al.)," ICC Summary for Policymakers, January 25, 2020, iccwbo.org/content/uploads/sites/3/2021/01/icc-summary-for-policy makers-the-economic-case-for-global-vaccination.pdf.

61 Matt Grainger and Sarah Dransfield, "Rich nations vaccinating one person every second while majority of the poorest nations are yet to give a single dose," UNAIDS, March 10, 2021, www.unaids.org/en/resources /presscentre/featurestories/2021/march/20210310.

62 Marco Haffner et al., "The global economic cost."

63 The Independent Panel, "Second Report."

64 Cristina Krippahl, "Africa lags in COVID-19 vaccination drive," *Deutsche Welle*, January 22, 2021, www.dw.com/en/africa-lags-in-covid -19-vaccination-drive/a-56289170.

65 Ben Guarino and Carolyn Y. Johnson, "UK Approves Anti-Inflammatory Drugs to Treat Sickest Covid-19 Patients After Strong Results in Clinical Trial," *Washington Post*, January 9, 2021, www.washington post.com/health/2021/01/08/covid-arthritis-drugs-treatment.

# Notes

66 Debora MacKenzie, "Is stockpiling pandemic flu drugs shrewd or misguided?," *New Scientist*, April 10, 2014, www.newscientist.com/article/dn25397-is-stockpiling-pandemic-flu-drugs-shrewd-or-misguided/#ixzz6sKVcKrO9.

67 Debora MacKenzie, "Evidence that Tamiflu reduces deaths in pandemic flu," *New Scientist*, June 24, 2013, www.newscientist.com/article/dn23744-evidence-that-tamiflu-reduces-deaths-in-pandemic-flu.

68 S. G. Muthuri et al., "Impact of neuraminidase inhibitor treatment on outcomes of public health importance during the 2009–2010 influenza A (H1N1) pandemic: A systematic review and meta-analysis in hospitalized patients," *The Journal of Infectious Diseases* 207, no. 4 (November 2012): 553–563, doi.org/10.1093/infdis/jis726.

69 Valerie Vaughn et al., "74. Empiric Antibiotic Therapy and Community-onset Bacterial Co-infection in Patients Hospitalized with COVID-19: A Multi-hospital Cohort Study," *Open Forum Infectious Diseases* 7, Supplement 1 (2020): S167–S168, doi.org/10.1093/ofid/ofaa439.384.

70 WHO Regional Office for Europe, "Preventing the COVID-19 pandemic from causing an antibiotic resistance catastrophe," November 18, 2020, www.euro.who.int/en/health-topics/disease-prevention/antimicrobial-resistance/news/news/2020/11/preventing-the-covid-19-pandemic-from-causing-an-antibiotic-resistance-catastrophe.

71 Debora MacKenzie, "The war against antibiotic resistance is finally turning in our favour," *New Scientist*, January 16, 2019, www.newscientist.com/article/2190957-the-war-against-antibiotic-resistance-is-finally-turning-in-our-favour.

72 The Review on Antimicrobial Resistance (chaired by Jim O'Neill), "Antimicrobial resistance: Tackling a crisis for the health and wealth of nations," December 2014, amr-review.org/sites/default/files/AMR%20Review%20Paper%20-%20Tackling%20a%20 crisis%20for%20the%20health%20and%20wealth%20of%20nations_1.pdf.

73 Carla Marinucci, "Schwarzenegger: 'Shortsighted' for California to defund pandemic stockpile he built," *Politico*, March 31, 2020, www.politico.com/states/california/story/2020/03/31/schwarzenegger

-shortsighted-for-california-to-defund-pandemic-stockpile-he-built
-1269954.

74 International Labour Organization (UN), "COVID-19 and the world
of work," www.ilo.org/global/topics/coronavirus/impacts-and-responses
/WCMS_739049/lang—en/index.htm

75 Caroline Huber et al., "The economic and social burden of the 2014
Ebola outbreak in West Africa," *The Journal of Infectious Diseases* 218, Supplement 5 (October 2018), doi.org/10.1093/infdis/jiy213.

76 Moki Edwin Kindzeka, "Cameroon reports polio cases amid
COVID scare," *VOA News*, March 7, 2021, www.voanews.com/covid
-19-pandemic/cameroon-reports-polio-cases-amid-covid-scare.

77 David N. Durrheim et al., "A dangerous measles future looms
beyond the COVID-19 pandemic," *Nature Medicine* 27, no. 3 (March 2021):
360–361, doi.org/10.1038/s41591-021-01237-5.

78 WHO Regional Office for Europe, "Statement — Catastrophic
impact of COVID-19 on cancer care," February 4, 2021, www.euro.who
.int/en/media-centre/sections/statements/2021/statement-catastrop
hic-impact-of-covid-19-on-cancer-care.

79 Kate M. Mitchell et al., "The potential effect of COVID-19-related
disruptions on HIV incidence and HIV-related mortality among men who
have sex with men in the USA: A modelling study," *The Lancet HIV* 8, no. 4
(February 2021), doi.org/10.1016/S2352-3018(21)00022-9.

80 Sameera Al Tuwaijri, "Killer # 2: Disrupted health services
during COVID-19," World Bank Blogs, August 20, 2020, blogs.worldbank
.org/health/killer-2-disrupted-health-services-during-covid-19.

81 UNICEF, "Disruptions in health services due to COVID-19 'may
have contributed to an additional 239,000 child and maternal deaths in
South Asia' — UN report," www.unicef.org/press-releases/disruptions
-health-services-due-covid-19-may-have-contributed-additional
-239000.

82 Pamela H. Lai et al., "Characteristics associated with out-of-
hospital cardiac arrests and resuscitations during the novel
coronavirus disease 2019 pandemic in New York City," *JAMA
Cardiology* 5, no. 10 (January 2020): 1154, doi.org/10.1001/jamacardio
.2020.2488.

83 Alexandra B. Hogan et al., "Report 19—The potential impact of the COVID-19 epidemic on HIV, TB and malaria in low- and middle-income countries," Imperial College London, May 1, 2020, www.imperial.ac.uk /mrc-global-infectious-disease-analysis/covid-19/report-19-hiv -tb-malaria.

84 The Global Fund, "Mitigating the impact of COVID-19 on countries affected by HIV, tuberculosis and malaria," June 2020, www.the globalfund.org/media/9819/covid19_mitigatingimpact_report_en.pdf.

85 Media brief, "One year on, new data show global impact of Covid-19 on TB epidemic is worse than expected," Stop TB Partnership, March 18, 2021, www.stoptb.org/webadmin/cms/docs/20210316_TB%20and%20 COVID_One%20Year%20on_Media%20Brief_FINAL.pdf.

86 "Transcript of October 2020 World Economic Outlook Press Briefing," International Monetary Fund, October 13, 2020, www.imf.org/en /News/Articles/2020/10/13/tr101320-transcript-of-october-2020-world -economic-outlook-press-briefing.

87 MacKenzie, "Plague! How to prepare for the next pandemic."

88 Congressional Budget Office, "Projected costs of U.S. nuclear forces, 2019 to 2028," January 24, 2019, www.cbo.gov/publication/54914.

89 World Health Organization, "Programme budget 2020–2021," 2019, www.who.int/about/finances-accountability/budget/en.

## CHAPTER 7: THINGS FALL APART

1 "The Most Important Jobs T-Shirt," Red Molotov, www.redmolotov .com/important-jobs-tshirt.

2 Debora MacKenzie, "Will a pandemic bring down civilisation?," *New Scientist*, April 2, 2008, www.newscientist.com/article/mg19826501 -400-will-a-pandemic-bring-down-civilisation.

3 Debora MacKenzie, "Why the demise of civilisation may be inevitable," *New Scientist*, April 2, 2008, www.newscientist.com/article/mg1982 6501-500-why-the-demise-of-civilisation-may-be-inevitable.

4 Thomas Homer-Dixon, "Complexity science," *Oxford Leadership Journal* 2, no. 1 (January 2011), homerdixon.com/complexity-science.

5 Edward N. Lorenz, "Predictability: Does the flap of a butterfly's wings in Brazil set off a tornado in Texas?," American Association for the

Advancement of Science, 139th meeting, December 29, 1972, eaps4.mit .edu/research/Lorenz/Butterfly_1972.pdf.

6 Thin Lei Win and Kim Harrisberg, "Africa faces 'hunger pandemic' as coronavirus destroys jobs and fuels poverty," Reuters, April 24, 2020, www .reuters.com/article/us-health-coronavirus-africa-hunger-feat/africa -faces-hunger-pandemic-as-coronavirus-destroys-jobs-and-fuels-pov erty-idUSKCN22629V.

7 "Acute hunger set to soar in over 20 countries, warn FAO and WFP," Food and Agriculture Organization of the United Nations, March 23, 2021, www.fao.org/news/story/en/item/1382490/icode.

8 Marius Gilbert et al., "Preparedness and vulnerability of African countries against importations of COVID-19: A modelling study," *The Lancet* 395, no. 10227 (March 2020): 871–877, doi.org/10.1016/s0140-6736 (20)30411-6.

9 UK Government, "NERVTAG Update Note on B.1.1.7 Severity," assets .publishing.service.gov.uk/government/uploads/system/uploads/attach ment_data/file/961042/S1095_NERVTAG_update_note_on_B.1.1.7_se verity_20210211.pdf.

10 Scientific Pandemic Influenza Group on Modelling, "SPI-M Modelling Summary," November 2018, assets.publishing.service.gov.uk /government/uploads/system/uploads/attachment_data/file/756738 /SPI-M_modelling_summary_final.pdf.

11 Civil Contingencies Secretariat (UK), "Preparing for pandemic influenza: Guidance for local planners," July 2013, assets.publishing.ser vice.gov.uk/government/uploads/system/uploads/attachment_data /file/225869/Pandemic_Influenza_LRF_Guidance.pdf.

12 "Expert reaction to preprint on COVID-19 and patient-derived mutations," *Science Media Centre*, April 21, 2020, www.sciencemediacentre.org /expert-reaction-to-preprint-on-covid-19-and-patient-derived -mutations.

13 Ralph S. Baric, "Emergence of a highly fit SARS-CoV-2 variant," *New England Journal of Medicine* 383, no. 27 (December 2020): 2684–2686, doi .org/10.1056/nejmcibr2032888.

14 Philip Ball, "Variant B117: What we know about the new Covid mutation," *Prospect Magazine*, January 4, 2021, prospectmagazine.co.uk

/science-and-technology/new-strain-variant-covid-19-coronavirus
-lockdown-schools.

15 Public Health England, "SARS-CoV-2 variants of concern and variants under investigation in England," June 3, 2021, https://assets.publishing
.service.gov.uk/government/uploads/system/uploads/attachment_data
/file/991343/Variants_of_Concern_VOC_Technical_Briefing_14.pdf.

16 Adam Kucharski, Twitter thread (5 tweets), December 28, 2020,
threadreaderapp.com/thread/1343567425107881986.html.

17 Andrew F. Read et al., "Imperfect vaccination can enhance the transmission of highly virulent pathogens," *PLoS Biology* 13, no. 7 (July 2015), doi.org/10.1371/journal.pbio.1002198.

18 Eric J. Hass et al., "Impact and effectiveness of mRNA BNT162b2 vaccine against SARS-CoV-2 infections and COVID-19 cases, hospitalisations, and deaths following a nationwide vaccination campaign in Israel: an observational study using national surveillance data," *The Lancet* 397, no. 10287 (May 2021):1819–1829, doi.org/10.1016/S0140-6736(21)00947-8.

19 Leen Vijgen et al., "Complete genomic sequence of human Coronavirus OC43: Molecular clock analysis suggests a relatively recent zoonotic Coronavirus transmission event," *Journal of Virology* 79, no. 3 (January 2005): 1595–1604, doi.org/10.1128/jvi.79.3.1595-1604.2005.

20 Jennie S. Lavine, Ottar N. Bjornstad, and Rustom Antia, "Immunological characteristics govern the transition of COVID-19 to endemicity," *Science*, February 12, 2021, doi.org/10.1126/science.abe6522.

21 William E. Diehl et al., "Ebola virus glycoprotein with increased infectivity dominated the 2013–2016 epidemic," *Cell* 167, no. 4 (November 2016): 1088–1098.e6, doi.org/10.1016/j.cell.2016.10.014.

22 MacKenzie, "Ebola rapidly evolves to be more transmissible and deadlier."

23 Ed Feil and Christian Yates, "Will coronavirus really evolve to become less deadly?," *The Conversation*, February 1, 2021, theconversa
tion.com/will-coronavirus-really-evolve-to-become-less-deadly
-153817.

24 Peter Kerr et al., "Myxoma virus and The leporipoxviruses: An evolutionary paradigm," *Viruses* 7, no. 3 (March 2015): 1020–1061, doi
.org/10.3390/v7031020.

25 Alan Mckinnon, "Life without trucks: The impact of a temporary disruption of road freight transport on a national economy," *Journal of Business Logistics* 27, no. 2 (May 2006): 227–250, doi.org/10.1002/j.2158 -1592.2006.tb00224.x.

26 MacKenzie, "Will a pandemic bring down civilisation?"

27 Department for Business, Energy, and Industrial Strategy, and Health and Safety Executive (UK government), "Guidance: Preparing for and responding to energy emergencies," January 9, 2020, www.gov.uk /guidance/preparing-for-and-responding-to-energy-emergencies.

28 Department of Energy and Climate Change (UK), "DECC approach to dealing with pandemic illness in the upstream energy sector," July 24, 2013, assets.publishing.service.gov.uk/government/uploads/system/uplo ads/attachment_data/file/48946/Dealing_with_pandemic_illness_in _the_upstream_energy_sector.doc.

29 Cybersecurity and Infrastructure Security Agency (US Department of Homeland Security), "Guidance on the essential critical infrastructure workforce," April 24, 2020, www.cisa.gov/publication/guidance -essential-critical-infrastructure-workforce.

30 Xiao Huang, "Staying at Home Is a Privilege: Evidence from Fine-Grained Mobile Phone Location Data in the United States during the COVID-19 Pandemic," *Annals of the American Association of Geographers*, May 27, 2021, https://doi.org/10.1080/24694452.2021.1904819.

31 The OpenSAFELY Collaborative et al., "OpenSAFELY: Factors associated with COVID-19-related hospital death in the linked electronic health records of 17 million adult NHS patients," medRxiv, May 7, 2020, doi.org/10.1101/2020.05.06.20092999.

32 Lena Masri, "COVID-19 takes unequal toll on immigrants in Nordic region," Reuters, April 24, 2020, www.reuters.com/article/us-health -coronavirus-norway-immigrants-idUSKCN2260XW.

33 Samrachana Adhikari et al., "Assessment of community-level disparities in Coronavirus disease 2019 (COVID-19) infections and deaths in large US metropolitan areas," *JAMA Network Open* 3, no. 7 (July 2020), doi .org/10.1001/jamanetworkopen.2020.16938.

34 Denis Campbell and Caroline Bannock, "Up to 100 UK children a week hospitalised with rare post-Covid disease," *The Guardian*,

February 5, 2021, www.theguardian.com/world/2021/feb/05/up-to-100-uk -children-a-week-hospitalised-with-rare-post-covid-disease.

35 Office for National Statistics (UK government), "Updating ethnic contrasts in deaths involving the coronavirus (COVID-19), England and Wales: Deaths occurring 2 March to 28 July 2020," October 16, 2020, www.ons .gov.uk/peoplepopulationandcommunity/birthsdeathsandmarriages /deaths/articles/updatingethniccontrastsindeathsinvolvingthecorona viruscovid19englandandwales/deathsoccurring2marchto28july2020.

36 Shirley Sze et al., "Ethnicity and clinical outcomes in COVID-19: A systematic review and meta-analysis," *The Lancet* 29, no. 100630 (November 2020), doi.org/10.1016/j.eclinm.2020.100630.

37 Benjamin D. Renelus et al., "Racial disparities in COVID-19 hospitalization and in-hospital mortality at the height of the New York City pandemic," *Journal of Racial and Ethnic Health Disparities* (December 2020): e2026881, doi.org/10.1007/s40615-020-00872-x.

38 "Stranded seafarers: A 'humanitarian crisis,'" International Labour Organization, September 15, 2020, www.ilo.org/global/about-the-ilo/news room/news/WCMS_755390/lang—en/index.htm.

39 BBC, "Coronavirus: Lack of co-ordination let virus spread—UN's Guterres," television newscast, interview by Nick Bryant, May 1, 2020, www.bbc.com/news/av/world-us-canada-52496983/coronavirus-lack -of-co-ordination-let-virus-spread-un-s-guterres.

40 Warner Raza et al., "Post Covid-19 value chains: Options for reshoring production back to Europe in a globalised economy," European Parliament Think Tank, February 19, 2021, doi.org/10.2861/118324.

41 Shannon K. O'Neill, "How to pandemic-proof globalization," *Foreign Affairs*, April 1, 2020, www.foreignaffairs.com/articles/2020-04-01/how -pandemic-proof-globalization.

42 Richard Webb, Debora MacKenzie, and John Horgan, "What happens when society crumbles and progress stops," *New Scientist*, June 1, 2016, www.newscientist.com/article/mg23030760-500-the-end-what-hap pens-when-society-crumbles-and-progress-stops.

43 Adele Berti, "The impact of Covid-19 on global shipping: Part 1, system shock," *Ship Technology*, April 2, 2020, www.ship-technology.com /features/impact-of-covid-19-on-shipping.

44 Thomas Homer-Dixon et al., "Synchronous failure: The emerging causal architecture of global crisis," *Ecology and Society* 20, no. 3 (2015), doi.org/10.5751/es-07681-200306.

## CHAPTER 8: THE PANDEMIC THAT NEVER SHOULD HAVE HAPPENED — AND HOW TO STOP THE NEXT ONE

1 Sara Frueh, "NAS annual meeting: Experts discuss COVID-19 pandemic and science's response," The National Academies of Science and Engineering, April 27, 2020, www.nationalacademies.org/news/2020/04 /nas-annual-meeting-experts-discuss-covid-19-pandemic-and-sciences -response.

2 John F. Kennedy, "Remarks at the Convocation of the United Negro College Fund, Indianapolis, Indiana, April 12, 1959," JFK Library, www .jfklibrary.org/archives/other-resources/john-f-kennedy-speeches/india napolis-in-19590412. The quote is slightly different in its other iteration from October 1960.

3 Pekar et al., "Timing the SARS-CoV-2 index case in Hubei Province."

4 Zeynep Tufekci, "How the coronavirus revealed authoritarianism's fatalflaw," *TheAtlantic*,February22,2020,www.theatlantic.com/technology /archive/2020/02/coronavirus-and-blindness-authoritarianism /606922.

5 John Garrick and Yan Bennett, "How China is controlling the COVID origins narrative."

6 Kynge, Yu, and Hancock, "Coronavirus: The cost of China's public health cover-up."

7 Debora MacKenzie, "Can we afford not to track deadly viruses?," *New Scientist*, May 20, 1995, www.newscientist.com/article/mg14619780 -300-can-we-afford-not-to-track-deadly-viruses.

8 Nicholas Kulish, Sarah Kliff, and Jessica Silver-Greenberg, "The U.S. tried to build a new fleet of ventilators. The mission failed," *New York Times*, March 29, 2020, www.nytimes.com/2020/03/29/business/corona virus-us-ventilator-shortage.html.

9 Kristian Andersen, "nCoV-2019 codon usage and reservoir (not snakes v2)," *Virological*, January 24, 2020, virological.org/t/ncov-2019-codon -usage-and-reservoir-not-snakes-v2/339.

# Notes

10 Natalie Berkhout, "China to end sale of live poultry at wet markets," *Poultry World*, July 10, 2020, www.poultryworld.net/Meat/Articles/2020/7/China-to-end-sale-of-live-poultry-at-wet-markets-611401E.

11 Jane Qiu, "How China's 'Bat Woman' hunted down viruses from SARS to the new Coronavirus," *Scientific American*, April 27, 2020, www.scientificamerican.com/article/how-chinas-bat-woman-hunted-down-viruses-from-sars-to-the-new-coronavirus1.

12 Kristian G. Andersen et al., "The proximal origin of SARS-CoV-2."

13 G20, "G20 leaders' statement, extraordinary G20 leaders' summit statement on COVID-19," March 26, 2020, g20.org/en/media/Documents/G20_Extraordinary%20G20%20Leaders%E2%80%99%20Summit_Statement_EN%20(3).pdf.

14 "SARS-CoV-2 variant classifications and definitions," Centers for Disease Control and Prevention, updated March 24, 2021, www.cdc.gov/coronavirus/2019-ncov/cases-updates/variant-surveillance/variant-info.html.

15 Mike Stobbe, "Health official says US missed some chances to slow virus," Associated Press, May 1, 2020, apnews.com/a758f05f337736e93dd0c280deff9b10.

16 Gary P. Pisano, Raffaella Sadun, and Michele Zanini, "Lessons from Italy's response to coronavirus," *Harvard Business Review*, March 27, 2020, hbr.org/2020/03/lessons-from-italys-response-to-coronavirus.

17 Adam Tooze, "How coronavirus almost brought down the global financial system," *Guardian*, April 14, 2020, www.theguardian.com/business/2020/apr/14/how-coronavirus-almost-brought-down-the-global-financial-system.

18 Christopher J. Fettweis, "Unipolarity, hegemony, and the new peace," *Security Studies* 26, no. 3 (August 2017): 423–451, doi.org/10.1080/09636412.2017.1306394.

19 Council on Foreign Relations, "U.S. must apply lessons from 'deeply flawed' pandemic response to preempt a deadlier disaster, warns CFR Task Force," October 8, 2020, www.cfr.org/news-releases/us-must-apply-lessons-deeply-flawed-pandemic-response-preempt-deadlier-disaster-warns#:~:text=%E2%80%9CAs%20in%20other%20recent%20outbreaks,international%20concern%20over%20China's%20objections.%E2%80%9D.

# Notes

20 Lawrence Gostin and Sarah Wetter, "Two legal experts explain why the U.S. should not pull funding from the WHO amid COVID-19 pandemic."

21 BBC, "Coronavirus: Lack of co-ordination let virus spread—UN's Guterres."

22 Debora MacKenzie, "World must get ready now for the next big health threat."

23 Anne-Marie Slaughter, *The Chessboard and the Web: Strategies of Connection in a Networked World* (New Haven, CT: Yale University Press, 2017).

24 "Independent Task Force Report No. 78: Improving Pandemic Preparedness: Lessons from COVID-19," Council on Foreign Relations, October 2020, www.cfr.org/report/pandemic-preparedness-lessons-COVID-19 /pdf/TFR_Pandemic_Preparedness.pdf.

25 Maria Espona, "Early warning mechanism in case of alleged use," in *Verifying the BTWC in a Fast-Changing World*, ed. J. P. Zanders (London: World Scientific Publishing, 2021).

26 Peter Beaumont and Julian Borger, "WHO warned of transmission risk in January, despite Trump claims," *The Guardian*, April 9, 2000, www.theguardian.com/world/2020/apr/09/who-cited-human-trans mission-risk-in-january-despite-trump-claims.

27 The Independent Panel, "Second Report."

28 "Independent Task Force Report No. 78."

29 WHO, "COVID-19 shows why united action is needed for more robust international health architecture," March 30, 2021, www.who.int /news-room/commentaries/detail/op-ed---covid-19-shows-why -united-action-is-needed-for-more-robust-international-health-arch itecture.

30 Haik Nikogosian, "How would a pandemic treaty relate with the existing IHR (2005)?," *BMJ Opinion*, May 23, 2021, https://blogs. bmj.com/bmj/2021/05/23/how-would-a-pandemic-treaty-relate-with -the-existing-ihr-2005/.

31 Debora MacKenzie, "US may respond after chemical weapons attack in Syria," *New Scientist*, April 11, 2018, www.newscientist.com /article/mg23831733-600-us-may-respond-after-chemical-weapons -attack-in-syria.

# Notes

32 Organisation for the Prohibition of Chemical Weapons, "Chemical Weapons Convention," September 27, 2005 (revised), www.opcw.org /chemical-weapons-convention.

33 Jonathan B. Tucker, "The chemical weapons convention: Has it enhanced U.S. security?," *Arms Control Today*, April 2001, www.arms control.org/act/2001-04/features/chemical-weapons-convention -enhanced-us-security.

34 World Health Organization, "Global Polio Eradication Initiative," polioeradication.org.

35 Matthew H. Bonds et al., "Poverty trap formed by the ecology of infectious diseases," *Proceedings of the Royal Society B: Biological Sciences* 277, no. 1685 (2009): 1185–1192, doi.org/10.1098/rspb.2009.1778.

36 Debora MacKenzie, "The great flu cover-up," *New Scientist*, January 31, 2004, www.newscientist.com/article/mg18124320-200-the-great -flu-cover-up.

37 Debora MacKenzie, "Chasing deadly viruses for a living," *New Scientist*, July 4, 2012, www.newscientist.com/article/mg21528722-100-chas ing-deadly-viruses-for-a-living.

38 António Guterres, "Secretary-General's remarks at G-20 virtual summit on the COVID-19 pandemic," United Nations, March 26, 2020, www .un.org/sg/en/content/sg/statement/2020-03-26/secretary-generals -remarks-g-20-virtual-summit-the-covid-19-pandemic.

39 Tim Walker, Twitter post, March 28, 2020, 2:03 p.m., twitter.com /ThatTimWalker/status/1243961867116204032.

40 Maitreesh Ghatak, Xavier Jaravel, and Jonathan Weigel, "The world has a $2.5 trillion problem. Here's how to solve it," *New York Times*, April 20, 2020, www.nytimes.com/2020/04/20/opinion/coronavirus-economy -bailout.html.

41 Scott Gottlieb, "The CIA Can Help Spot the Next Pandemic," *Wall Street Journal*, February 28, 2021, www.wsj.com/articles/the-cia -can-help-spot-the-next-pandemic-11614548072.

42 Mark Schaller, "The behavioural immune system and the psychology of human sociality," *Philosophical Transactions of the Royal Society B: Biological Sciences* 366, no. 1583 (December 2011): 3418–3426, doi .org/10.1098/rstb.2011.0029.

# Notes

43 Kathleen McAuliffe, "Liberals and conservatives react in wildly different ways to repulsive pictures," *The Atlantic*, March 2019, www.theat lantic.com/magazine/archive/2019/03/the-yuck-factor/580465.

44 Corinne J. Brenner and Yoel Inbar, "Disgust sensitivity predicts political ideology and policy attitudes in the Netherlands," *European Journal of Social Psychology* 45, no. 1 (November 2014): 27–38, doi.org/10.1002/ejsp.2072.

45 Corey L. Fincher et al., "Pathogen prevalence predicts human cross-cultural variability in individualism/collectivism," *Proceedings of the Royal Society B: Biological Sciences* 275, no. 1640 (February 2008): 1279–1285, doi.org/10.1098/rspb.2008.0094.

46 Debora MacKenzie, "How your personality predicts your attitudes towards Brexit," *New Scientist*, July 9, 2018, www.newscientist.com/article/2173681-how-yourpersonality-predicts-your-attitudes-towards-brexit.

47 Leor Zmigrod et al., "The psychological and socio-political consequences of infectious diseases," *PxyArXiv Preprints* (April 11, 2020), doi.org/10.31234/osf.io/84qcm.

48 Sarah Estes and Jesse Graham, "Why Gay Marriage Divides the World," *New Scientist*, May 16, 2012, www.newscientist.com/article/mg21428655-700-why-gay-marriage-divides-the-world.

49 Alec T. Beall et al., "Infections and elections," *Psychological Science* 27, no. 5 (March 14, 2016): 595–605, doi.org/10.1177/0956797616628861.

50 Philip Bump, "Donald Trump's lengthy and curious defense of his immigrant comments, annotated," *Washington Post*, July 6, 2015, www.washingtonpost.com/news/the-fix/wp/2015/07/06/donald-trumps-lengthy-and-curious-defense-of-his-immigrant-comments-annotated.

51 Timothy W. Russell et al., "Effect of internationally imported cases on internal spread of COVID-19: A mathematical modelling study," *The Lancet Public Health* 6, no. 1 (January 2021), doi.org/10.1016/s2468-2667(20)30263-2.

52 Dan Cable and Francesca Gino, "Coping with 'death awareness' in the COVID-19 era," *Scientific American*, May 13, 2020, www.scientificameri can.com/article/coping-with-death-awareness-in-the-covid-19-era.

53 Tom Pyszczynski et al., "Terror management theory and the COVID-19 pandemic," *Journal of Humanist Psychology* 0022167820959488 (September 2020), doi.org/10.1177/0022167820959488.

54 Michael Matthews, "Mortality salience during a pandemic," *Psychology Today*, September 4, 2020, www.psychologytoday.com/us/blog /head-strong/202009/mortality-salience-during-pandemic.

55 Tierra Smiley Evans et al., "Synergistic China-US ecological research is essential for global emerging infectious disease preparedness," *EcoHealth* 17, no. 1 (March 2020): 160–173, doi.org/10.1007/s10393-020-01471-2.

56 Rebecca Solnit, *A Paradise Built in Hell* (New York: Viking, 2009).

57 Steven Lee Myers, "China spins tale that the U.S. Army started the coronavirus epidemic," *New York Times*, March 13, 2020, www.nytimes .com/2020/03/13/world/asia/coronavirus-china-conspiracy-theory .html. Other examples include Emma Graham-Harrison and Robin McKie's "A year after Wuhan alarm, China seeks to change Covid origin story" (*The Guardian*), and Javier C. Hernández's "China Peddles Falsehoods to Obscure Origin of Covid Pandemic" (*New York Times*).

58 Marc A. Thiessen, "China should be legally liable for the pandemic damage it has done," *Washington Post*, April 9, 2020, www.washing tonpost.com/opinions/2020/04/09/china-should-be-legally-liable-pan demic-damage-it-has-done.

59 "Statement: Saving Lives in America, China, and Around the World," signed Madeleine Albright et al., UC San Diego 21 Century China Center, April 3, 2020, china.ucsd.edu/_files/statement/covid-19-pande mic-statement.pdf.

60 Laurens Cerulus, "Ursula von der Leyen backs probe into how coronavirus emerged," *Politico EU*, May 1, 2020, politico.eu/article/von -der-leyen-backs-probe-into-how-coronavirus-emerged.

61 The version quoted here is a slightly refined version Farrar tweeted the day after the talk: Jeremy Farrar, Twitter post, April 26, 2020, 6:26 a.m., twitter.com/JeremyFarrar/status/1254356097470738432. For the original speech: Jeremy Farrar, "COVID-19 Update," panel discussion, National Academy of Sciences 157th Annual Meeting, April 25, 2020, online, www .nasonline.org/about-nas/events/annual-meeting/nas157/covid19 -update.html.